Fundamentals of Modern Manufacturing: Materials, Processes and Systems

Fundamentals of Modern Manufacturing: Materials, Processes and Systems

Hunter Bale

𝓒𝓛 LANRYE
INTERNATIONAL
www.clanryeinternational.com

Clanrye International,
750 Third Avenue, 9th Floor,
New York, NY 10017, USA

ISBN: 978-1-64726-650-9

Cataloging-in-Publication Data

Fundamentals of modern manufacturing : materials, processes and systems / Hunter Bale.
p. cm.
Includes bibliographical references and index.
ISBN 978-1-64726-650-9
1. Production engineering. 2. Manufacturing processes.
3. Manufacturing industries--Technological innovations.
I. Bale, Hunter.
TS183 .M63 2023
670--dc23

For information on all Clanrye International publications
visit our website at www.clanryeinternational.com

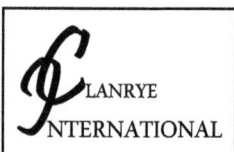

𝒞LANRYE
INTERNATIONAL

Contents

Preface

Manufacturing refers to the utilization of machinery, tools, chemical processing, and human labor to transform raw materials or components into finished goods. It can be broadly categorized into contract manufacturing, additive manufacturing, and advanced manufacturing. Modern manufacturing is central to industrial production, from the production of base materials to the production of semi-finished goods and finished goods. A number of cutting-edge techniques have been developed over the past years that enable manufacturing processes to be more environment friendly, adaptable and less energy-consuming. This book contains some path-breaking studies on modern manufacturing as well as the materials, processes and systems involved in it. It will also provide interesting topics for research which interested readers can take up. The book will help the readers in keeping pace with the rapid changes in this field.

The information contained in this book is the result of intensive hard work done by researchers in this field. All due efforts have been made to make this book serve as a complete guiding source for students and researchers. The topics in this book have been comprehensively explained to help readers understand the growing trends in the field.

I would like to thank the entire group of writers who made sincere efforts in this book and my family who supported me in my efforts of working on this book. I take this opportunity to thank all those who have been a guiding force throughout my life.

Hunter Bale

Theory of Metal Cutting

1.1   Single Point Cutting Tool Nomenclature and Geometry

This type of tool has a effective cutting edge and removes excess material from the work piece along the cutting edge. Single point cutting tool is of the following types:

- Ground type

- Tipped type

- Forged type

- Bit type

In ground type the cutting edge is formed by grinding the end of a piece of tool steel stock. Whereas in forged type the cutting edge is formed by rough forging before hardening and grinding. In tipped type cutting tool the cutting edge is in the form of a small tip made of high grade material which is welded to a shank made up of lower grade material. In bit type, a high grade material of a square, rectangular or some other shape is held mechanically in a tool holder.

Single point tools are commonly used in lathes, shapers, planers, boring machines and slotters. Single point cutting tool may be left handed or right handed type. A tool is said to be right/ left hand type if the cutting edge is on the right or left side when viewing tool from the point end lathe tools, shaper tools, planner tool and boring tools are single point tools.

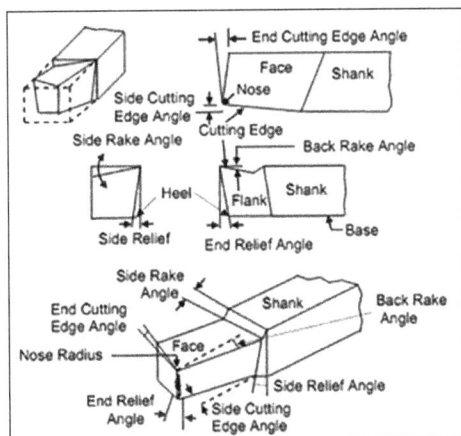

Nomenclature of single point cutting tool.

Geometry of Single Point Cutting Tool

- Shank: It is the main body of the tool.

- Base: The portion of the shank that lies opposite to the top face of the shank is called base.

The single point cutting tool has two edges and these are:

- End cutting edge angle: End cutting edge angle is the angle between the end cutting edge and a line perpendicular to the shank of the tool. It provides clearance between tool cutting edge and work piece.

- Side cutting edge angle: Side cutting edge angle is the angle between straight cutting edge on the side of tool and the side of the shank. It is responsible for turning the chip away from the finished surface.

- Flank: The surface or surfaces adjacent to the cutting edge is called flank of the tool.

- Heel: It is a curved portion and intersection of the base and flank of the tool.

- Cutting Edge: It is the edge on the face of the tool which removes the material from the work piece. The cutting edge consists of the side cutting edge (major cutting edge) and cutting edge(minor cutting edge) and the nose.

- Nose radius: It is the radius of the nose. Nose radius increases the life of the tool and provides better surface finish.

- Back rake angle: Back rake angle is the angle between the face of the single point cutting tool and a line parallel with base of the tool measured in a perpendicular plane through the side cutting edge. If the slope face is downward toward the nose, it is negative back rake angle and if it is upward toward nose, it is positive back rake angle. Back rake angle helps in removing the chips away from the work piece.

- End relief angle: End relief angle is defined as the angle between the portion of the end flank immediately below the cutting edge and a line perpendicular to the base of the tool, measured at right angles to the flank. End relief angle allows the tool to cut without rubbing on the work piece.

- Side rake angle: Side rake angle is the angle by which the face of tool is inclined side ways. Side rake angle is the angle between the surface the flank immediately below the point and the line down from the point perpendicular to the base. Side rake angle of cutting tool determines the thickness of the tool behind the cutting edge. It is provided on tool to provide clearance between work piece and tool so as to prevent the rubbing of work piece with end flake of tool.

- Side relief angle: Side rake angle is the angle between the portion of the side flank immediately below the side edge and a line perpendicular to the base of the tool measured at right angles to the side. Side relief angle is the angle that prevents the interference as the tool enters the material. It is incorporated on the tool to provide relief between its flank and the work piece surface.

Types and Applications of Different Types of Cutting Tools

All cutting tools can be divided into two groups. They are:

- Single point tool

- Multi point tool

Single point cutting tools having a wedge like action, finds a wide application on lathes and slotting machines, etc.

Multi-point cutting tools are merely two or more single point tools arranged together as a unit.

Types of Single Point Lathe Tool

- According to the method of manufacturing of the tool:

  ○ Forged tool.

  ○ Tipped tool brazed to the carbon steel shank.

  ○ Tipped tool fastened mechanically to the carbons tool shank.

- According to the method of holding the tool:

  ○ Solid tool.

  ○ Tool bit inserted in the tool holder.

- According to the method of using the tool:

  ○ Turning.

  ○ Chamfering.

  ○ Thread cutting.

- Facing:

  ○ Grooving.

- ○ Forming.

- ○ Boring.

- ○ Internal thread cutting.

- ○ Parting off.

- According to the method of applying feed:

- ○ Right hand.

- ○ Left hand.

- ○ Round nose.

Types of Multi Point Tool

- Plain milling cutter.

- Side milling cutter.

- Metal slitting saw.

- Angle milling cutter.

- End milling cutter.

- T-slop milling cutter.

- Wood ruff key slot milling cutter.

- Fly cutter.

- Formed cutter.

- Tap and reamer cutter.

Application of Different Types of Cutting Tools

Operation	Application
1) Turning	Turning in a Lathe is to remove excess material from the work piece to produce a cone shaped or a cylindrical surface.
2) Chamfering	Chamfering is the operation of beveling the extreme end of a work piece. This is used to remove the burrs, to protect the end of the work piece from being damaged and to have a better look.
3) Facing	Facing is the operation of machining the ends of a piece of work to produce a flat surface.

4) Grooving	Grooving is the process of reducing the diameter of a work piece over a very narrow surface.
5) Boring	Boring is the operation of enlarging and truing a hole produced by drilling, punching, casting, etc.
6) End milling	The end mills are used for light milling operations like cutting slots, machining, Accurate holes, producing narrow flat surfaces and for profile milling operations.
7) Plain milling	Used to produce flat, horizontal surfaces.

1.2   Mechanics of Chip Formation

Machining is a semi-finishing or finishing process essentially done to impart required or stipulated dimensional and form accuracy and surface finish to enable the product to:

- Fulfill its basic functional requirements.
- Render long service life.
- Provide better or improved performance.

Machining is a process of gradual removal of excess material from the preformed blanks in the form of chips. The form of the chips is an important index of machining because it directly or indirectly indicates:

- Nature and behavior of the work material under machining condition.
- Specific energy requirement (amount of energy required to remove unit volume of work material) in machining work.

The form of machined chips depends mainly upon:

- Work material.
- Machining environment or cutting fluid that affects temperature and friction at the chip-tool and work-tool interfaces.
- Material and geometry of the cutting tool.
- The levels of cutting velocity and feed and also to some extent on depth of cut.
- Knowledge of basic mechanism(s) of chip formation helps to understand the characteristics of chips and to attain favorable chip forms.

In a figure the tool is considered stationary and the work piece moves to the right. The metal is mainly compressed in the area in front of the cutting tool. It will result in high temperature shear and plastic flow if the metal is ductile.

When the stress in the work piece just ahead of the cutting tool reaches a value exceeding the ultimate strength of the metal, particles will shear to form a chip element which moves up along the face of the work. The outward or shearing movement of both successive element is arrested by work hardening and the movement transferred is to the next element.

The process is repetitive and a continuous chip is formed having a maximum compressed and burnished underside and a minutely serrated top side caused by the shearing action. The place along which the element shears is known as shear plane. Thus the chip is formed by plastic deformation of the grain structure of the metal along the shear plane as shown in the figure below.

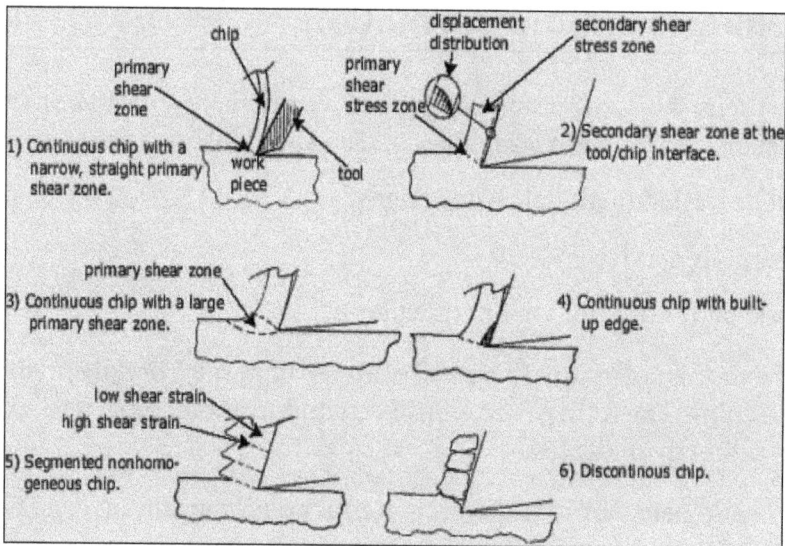

Structure of the metal along the shear plane.

The deformation does not occur sharply across the shear plane, but rather it occurs along a narrow band. The structure begins elongating along the line AB below the shear plane and continue to do so until it is completely deformed along the line CD above the shear plane in figure.

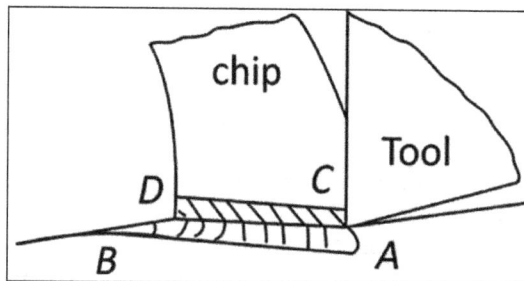

Shear zone during metal cutting.

The region between the lower surface AB, where elongation of the grain structure begins and the upper surface CD, where it is completed and the chip is born, is called the shear zone or primary deformation zone.

In figure the shear zone is included between two parallel lines AB and CD. Currently these two lines may not be parallel but may produce a wedge-shaped zone which is thicker near the tool face at the right than at the left. This is one of the causes of curling of chips in metal cutting.

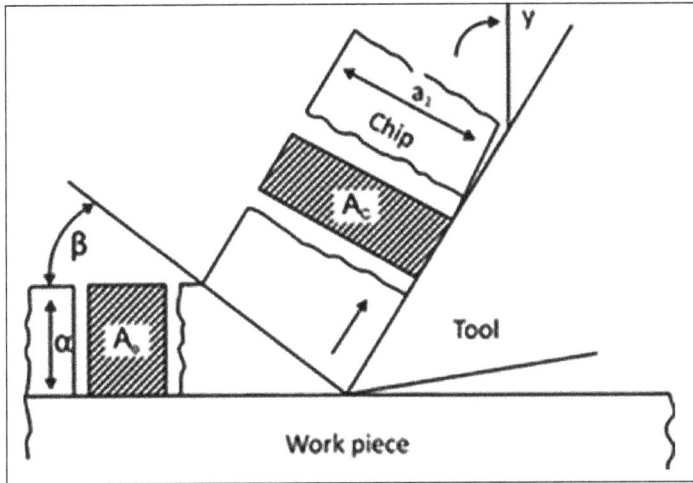

Geometry of chip formation in orthogonal cutting.

In addition, owing to the non-uniform distribution of forces at the chip-tool interface and on the shear plane the shear plane must be slightly curved concave downward. This also causes the chip to curl away from the cutting face of the tool.

Metal Removal in Metal Cutting

Basic elements of all the machining operations are as follows:

- Work piece
- Tool
- Chip

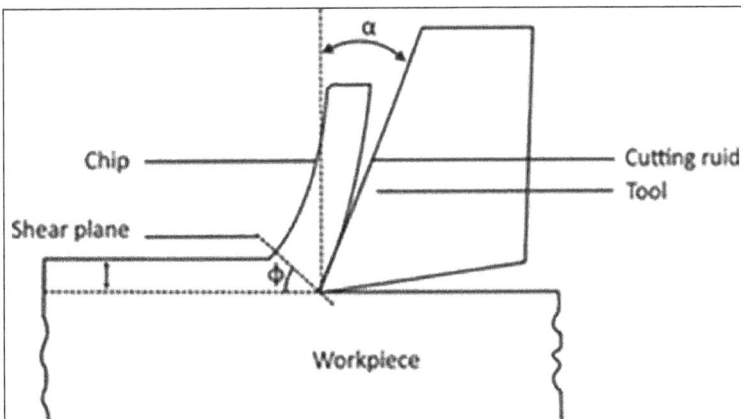

Basic elements of machining.

The figure shows the cutting action of a tool in a two dimensional or orthogonal position. To perform cutting operation, there should be relative motion between the tool and work piece. This relative motion is achieved by keeping either work piece stationary and moving the tool or tool stationary and moving the work piece. Sometimes, both are moving relative to each other.

The unwanted material is removed in the form of chip. The metal is severely compressed in the area in front of the cutting tool. This causes high temperature shear and plastic flow if the metal is ductile.

When the stress in the work piece just ahead the cutting tool reaches a value exceeding the ultimate strength of the metal, particles will shear to form a chip element which moves up along the face of the work. The plane along which the element shears is called the shear plane.

Mechanism of Chip Formation in Machining Brittle Materials

The basic two mechanisms involved in chip formation are:

- Yielding - generally for ductile materials.

- Brittle fracture - generally for brittle materials.

During machining, first a small crack develops at the tool tip as shown in the figure below, due to wedging action of the cutting edge. At the sharp crack-tip stress concentration takes place. In case of ductile materials immediately yielding takes place at the crack-tip and reduces the effect of stress concentration and prevents its propagation as crack. But in case of brittle materials the initiated crack quickly propagates, under stressing action and total separation takes place from the parent work piece through the minimum resistance path as indicated in the figure.

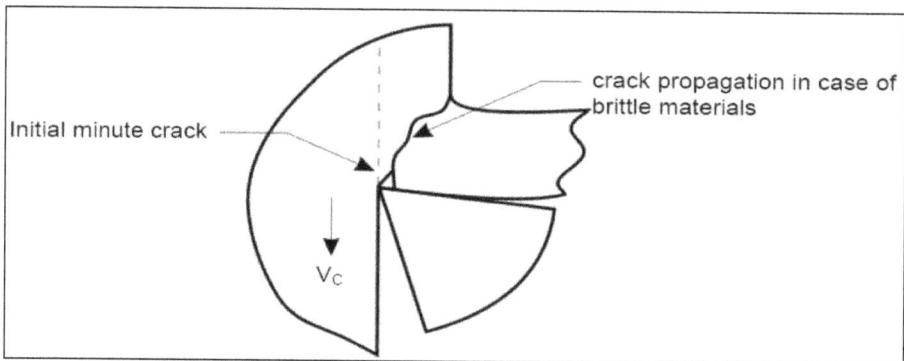

Development and propagation of crack causing chip separation.

Machining of brittle material produces discontinuous chips and mostly of irregular size and shape. The process of forming such chips is schematically shown in the figure below,(a, b, c, d and e).

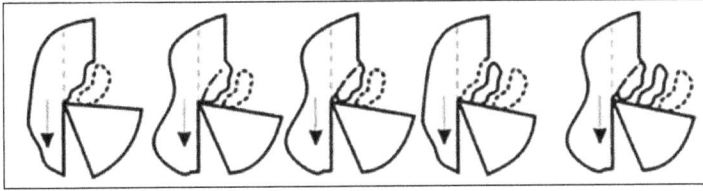

Schematic view of chip formation in machining brittle materials: (a) Separation (b) Swelling (c) Further swelling (d) Separation (e) Swelling again.

1.2.1 Types of Chips

The removed metal layer during the metal cutting operation is known as chips. The layer undergoes the following operations:

- Elastic deformation

- Plastic deformation

The type of chip depends on the material being cut and the cutting conditions:

- Continuous chip (without built up edge).

- Discontinuous chips.

- Continuous chips (with built up edge).

1. Continuous chip (without built up edge):

When the cutting tool moves towards the work piece, there occurs a plastic deformation of the work piece and the metal is separated without any discontinuity and it moves.

The chip moves along the face of the tool. This mostly occurs while cutting a ductile material. It is desirable to have smaller chip thickness and higher cutting speed in order to get continuous chips.

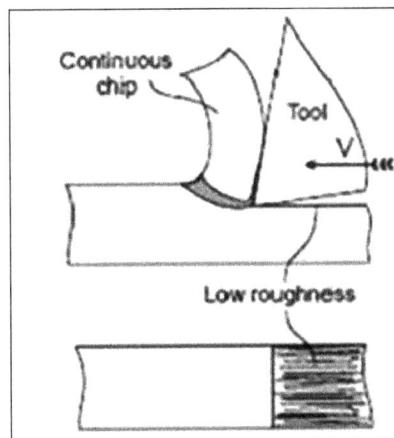

Continuous chip.

2. Discontinuous Chips:

This can also be known as segmental chips. This mostly occurs while cutting brittle material such as cast iron or low ductile materials. Instead of shearing the metal as it happens in the previous process, the metal is being fractured like segments of fragments and they pass over the tool faces.

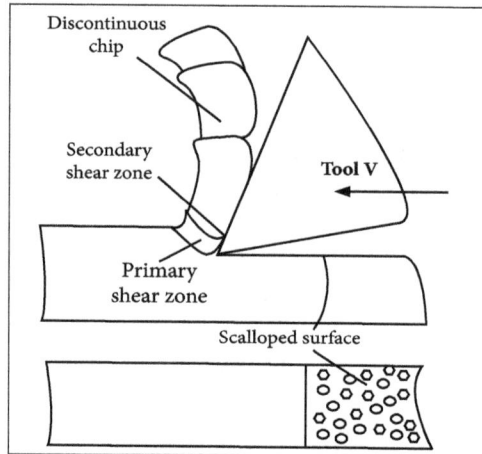

Discontinuous chips.

3. Continuous chips (with built up edge):

When cutting a ductile metal, the compression of the metal is followed by the high heat at tool face. This in turns enables part of the removed metal to be welded into the tool. This is known as built up edge.

1.3   Merchants Circle Diagram and Analysis

Assumptions made in drawing Merchant's circle:

- Shear surface is a plane extending upwards from the cutting edge.

- The tool is perfectly sharp and there is no contact along the clearance force.

- The cutting edge is a straight line extending perpendicular to the direction of motion and generates a plane surface as the work moves past it.

- The chip does not flow to either side, that is chip width is constant.

- The depth of cut remains constant.

- Width of the too, is greater than that of the work.

- Work moves with uniform velocity relative tool tip.

- No built up edge is formed.
  - The chip may be considered as a separate body held in equilibrium under the action of two equal, opposite and co linear resultant forces. i.e., F and F'.
  - The chip formation will be continuous without built up edge.
  - During cutting process, the cutting velocity remains constant.

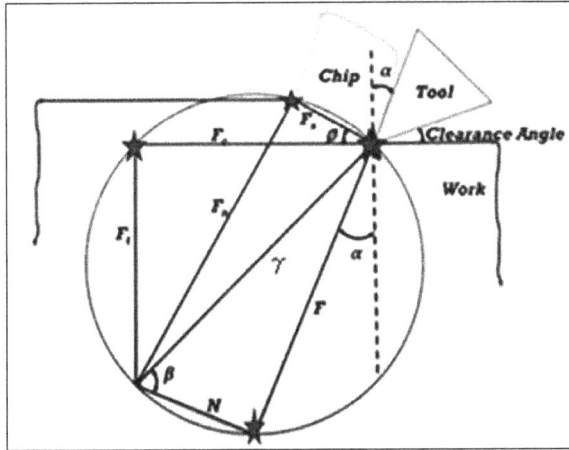

Merchant Circle.

In this diagram, two force triangles of figure have been combined and F_z and F_x together have been replaced by F. The circle has a diameter equal to F or F' passing through tool point as shown in figure.

The cutting force F_z and feed force F_s can be determined by using dynamo-meter. After measuring the F_z and F_x, they can be drawn into suitable scale. In the figure, let,

α = Rake angle of tool

β = Shear angle

γ = Friction angle

When the chip slides over the tool face under pressure, there may be some friction between these two. Therefore, the kinetic coefficient of friction (μ) can be expressed as coefficient of friction,

$$\mu = P / N = \tan \gamma$$

As mentioned earlier, the shear angle (β) can be obtained from the equation,

$$\tan \beta = r \cos\alpha / 1 - r \sin \alpha$$

Chip thickness ratio, $t = t_1 / t_2$

Where, t_1 and t_2 - Chip thickness before and after cutting respectively.

P-Frictional resistance

$P = F_x \cos \alpha + F_z \sin \alpha$

N = Normal force

$N = F_x \cos \alpha - F_z \sin \alpha$

F = Resultant force

Let, A_1 - Area of the chip before removal

A_s - Area of the shear plane

τ - Shear stress in the shear plane

$$\tau = \frac{\text{shear force}}{\text{Area}} = \frac{F_s}{A_s}$$

Shear stress,

$$A_s = \frac{A_1}{\sin \beta}$$

$$\therefore \tau = \frac{F_s}{A_1} \sin \beta$$

$$= \frac{(F_s \cos \beta - F_x \sin \beta) \sin \beta}{A_1}$$

Shear stress,

$$\tau = \frac{F_z \cos \beta \sin \beta - F_x \sin^2 \beta}{A_1}$$

F_z = Cutting force

$F_z = F \cos (\gamma - \alpha)$

$F_z = F \cos \theta$

Where, $\theta = \beta + \gamma - \alpha$

$$\therefore F = \frac{F_s}{\cos \theta}$$

Substituting F value in F_z equation we get,

$$\therefore F_z = \frac{F_s \cos(\gamma - \alpha)}{\cos\theta}$$

$$F_z = \frac{F_s \cos(\gamma - \alpha)}{\cos(\beta + \gamma - \alpha)} \quad [\theta = \beta + \gamma - \alpha]$$

We know that,

$$\mu = \tan\gamma = \frac{P}{N}$$

$$= \frac{F_x \cos\alpha + F_z \sin\alpha}{F_z \cos\alpha - F_x \sin\alpha}$$

The coefficient of friction,

$$\mu = \frac{F_x + F_z \tan\alpha}{F_z - F_x \tan\alpha}$$

The relation for F_s and F_n are as follows:

$$F_s = F_z \cos\beta - F_x \sin\beta$$
$$F_n = F_z \sin\beta + F_x \cos\beta$$

Advantageous use of Merchant's circle diagram enables the followings:

- Easy, quick and reasonably accurate determination of several other forces from a few known forces involved in machining.

- Friction at chip tool interface and dynamic yield shear strength can be easily determined.

- Equations relating the different forces are easily developed.

1.3.1 Ernst Merchant's Solution

This theory, first propagated by Ernst and Merchant in 1941, is based on the principle of minimum energy consumption. It implies that during cutting the metal shear should occur in that direction in which the energy requirement for shearing is minimum the other assumptions made by them include:

- The behavior of the metal being machined is like that of an ideal plastic.

- At the shear plane the shear stress is maximum is constant and independent of shear angle (ϕ).

They deduced the following relationship:

$$\phi = \frac{\pi}{4} - \frac{\beta}{2} + \frac{\gamma}{2}$$

Where,

ϕ= Shear angle

γ = Rake angle

β= Friction angle

Knowing F_c, F_t, α and ϕ, all other component forces can be calculated as:

$$F = F_c \sin \alpha + F_t \cos \alpha$$

$$N = F_c \cos \alpha - F_t \sin \alpha$$

The coefficient of friction will be then given as:

$$\mu = \frac{F}{N} = \frac{F_c \tan\alpha + F_t}{F_c - F_t \tan\alpha}$$

$$\lambda = \tan^{-1}\mu$$

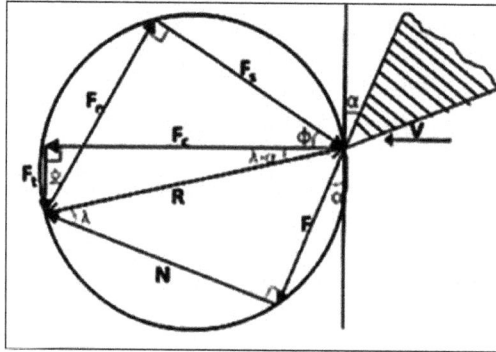

On shear plane,

$$F_s = F_c \cos\phi - F_t \sin\phi$$

$$F_n = F_c \sin\phi + F_t \cos\phi$$

Now,

$$F_t = F_n \cos\phi - F_s \sin\phi$$

$$F_c = F_n \sin\phi + F_s \cos\phi$$

Let, ϕ be the shear angle,

$$\tan\phi = \frac{r\cos\alpha}{1 - r\sin\alpha}$$

Where,

$$r = \frac{\text{Uncut chip thickness}}{\text{Chip thickness after cut}} = \frac{t}{t_c}$$

Now shear plane angle,

$$A_s = \frac{bt}{\sin\phi}$$

The average stresses on the shear plane area are,

$$\tau_S = \frac{F_S}{A_S}$$

$$\sigma_s = \frac{F_n}{A_n}$$

Now the shear force can be written as,

$$F_s = R\cos(\phi + \lambda - \alpha)$$

$$F_C = R\cos(\lambda - \alpha)$$

$$F_t = R\sin(\lambda - \alpha)$$

$$\tau_s = \frac{F_c \sec(\lambda - \alpha)\cos(\phi + \lambda - \alpha)\sin\phi}{bt}$$

Assuming that λ is independent of ϕ, for maximum shear stress,

$$\frac{\partial \tau_s}{\partial \phi} = 0$$

$$\cos(\phi + \lambda - \alpha)\cos\phi - \sin(\phi + \lambda - \alpha)\sin\phi = 0$$

$$\tan(\phi + \lambda - \alpha) = \cot\phi = \tan(90 - \phi)$$

$$\phi = 45° + \frac{\alpha}{2} - \frac{\lambda}{2}$$

1.3.2 Shear Angle Relationship, Problems of Merchant's Analysis

Shear angle (ϕ) can be measured by two methods:

- By measuring the thickness of the chip and the depth of cut.

- By taking a photo-micro-graph of the cutting process.

Back rake angle (α) of the tool can be determined by tool geometry.

Forces F_d and F_c could be measured with dynamometer. Therefore, once F_d, F_c, α and ϕ are known, all other components of forces acting on the chip can be determined with the geometry of Merchant's Circle Diagram.

To get the relationship, we draw the figure and relate the unknown forces with F_d and F_c.

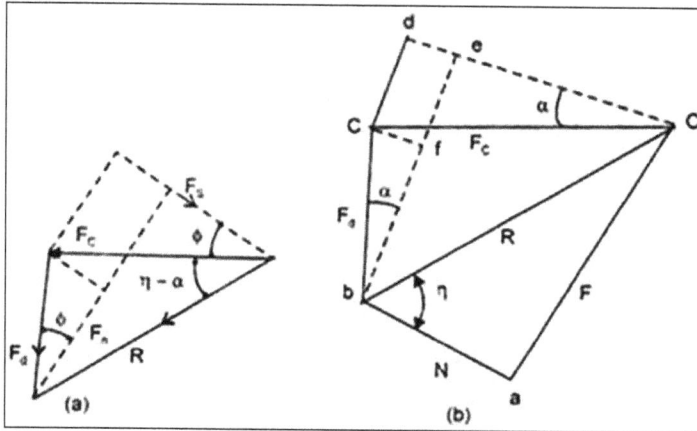

Cutting force vector diagram.

From the figure (a), we get,

$$F_s = F_c \cos\phi - F_d \sin\phi$$

$$F_n = F_c \sin\phi + F_d \cos\phi$$

$$= F_c \tan(\phi + \eta - \alpha)$$

$$F_s = R \cos(\eta - \alpha)$$

$$F_n = R \cos(\phi + \eta - \alpha)$$

Or, $R = \dfrac{F_x}{\cos(\phi + \eta - \alpha)}$

Therefore, $F_c = \dfrac{F_s}{\cos(\phi + \eta - \alpha)} \cos(\eta - \alpha)$ ...(i)

Or, $F_s = F_c \dfrac{\cos(\phi + \eta - \alpha)}{\cos(\eta - \alpha)}$ ...(ii)

From the figure (b), we get,

$$\text{Coefficient of friction} = \mu = \frac{F}{N} = \tan \eta \qquad ...(\text{iii})$$

Since, F= ao = be

    = ef +fb

    = cd+fb

$$F = F_c \sin\alpha + F_d \qquad ...(\text{iv})$$

Also N= ab = oe = od -de

$$N = F_c \cos\alpha - F_d \sin\alpha... \text{ v} \qquad ...(\text{v})$$

Now putting value of F and N in equation (iii), we get,

$$\mu = \frac{F}{N} = \frac{F_c \sin\alpha + F_d \cos\alpha}{F_c \cos\alpha + F_d \sin\alpha}$$

Dividing the numerator and denominator by cos α, we get,

$$\mu = \frac{F_c \tan\alpha + F_d}{F_c - F_d \tan\alpha} = \frac{F_d + F_c \tan\alpha}{F_c - F_d \tan\alpha}$$

$$\mu = \frac{F_d + F_c \tan\alpha}{F_c - F_d \tan\alpha}$$

The value of (μ) in metal cutting generally ranges for about 0.5 to 2. It indicates that, the chip encounters considerable frictional resistance while moving up the rake face of the tool.

If we make a free body analysis of the chip, forces acting on the chip would be as follows.

At cutting tool side due to motion of chip against tool there will be a frictional force and a normal force to support that, at material side thickness of the metal increases while it flows from uncut to cut portion.

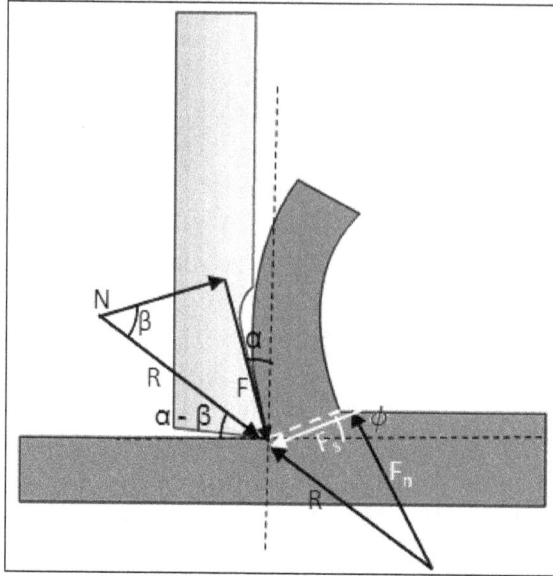

Forces acting on the chip on tool side and shear plane side.

This thickness increase is due to inter planar slip between different metal layers. There should be a shear force (F_s) to support this phenomenon. According to shear plane theory this metal layer slip happens at single plane called shear plane. So shear force acts on shear plane. Angle of shear plane can approximately determine using shear plane theory analysis. It is as follows:

$$\phi = 45° + \frac{\alpha}{2} - \frac{\beta}{2}$$

Shear force on shear plane can be determined using shear strain rate and properties of material. A normal force (F_n) is also present perpendicular to shear plane. The resultant force (R) at cutting tool side and metal side should balance each other in order to make the chip in equilibrium. Direction of resultant force, R is determined as shown in the figure above.

Problem

1. In an orthogonal cutting test with a tool of rake angle 10°, the following observations were made:

Chip thickness ratio = 0.3.

Horizontal component of cutting force = 1290 N.

Vertical component of cutting force = 1650 N.

From Merchant's theory, let us calculate the various components of the cutting forces and the coefficient of friction at the chip tool interface.

Solution:

Given Data:

- Rake angle $\alpha = 10°$.

- Chip thickness ratio $r = 0.3$.

- Horizontal cutting force $F_H = 1290$ N.

- Vertical cutting force $F_V = 1650$ N.

- Calculate F_S, N_S, F, N and μ.

(i) To find shear plane angle:

$$\tan\varphi = \frac{r\cos\alpha}{1 - r\sin\alpha}$$

$$= \frac{0.3 \times \cos 10°}{1 - 0.3 \times \sin 10°}$$

$$\tan\phi = \frac{0.295}{1 - 0.05209}$$

$$\phi = 17.29°$$

(ii) To find shear forces:

$$F_S = F_H \cos\phi \; F_V \sin\phi$$

$$= 1290 \times \cos 17.29 - [1650 \times \sin 17.29]$$

$$F_S = 1290 \times 0.9548 - 490.39$$

$$F_S = 741.3 \text{ N}$$

$$N_S = F_V \cos\phi + F_H \sin\phi$$

$$= (1650 \times \cos 17.29) + (1290 \times \sin 17.2)$$

$$N_S = 1958.8 \text{ N}$$

(iii) To find frictional forces:

$$F = F_H \sin\alpha + F_V \cos\alpha$$

$$= (1290 \times \sin 10°) + 1650 \times \cos 10°)$$

$F = 1848.9 \text{ N}$

$N = F_H \cos \alpha - F_V \sin \alpha$

$= (1290 \times \cos 10°) - (1650 \times \sin 10°)$

$N = 938.9 \text{ N}$

(iv) To find coefficient of friction:

$$\mu = \frac{F_V + F_H \tan \alpha}{F_H - F_V \tan \alpha}$$

$$= \frac{1650 + \left(1290 \times \tan 10°\right)}{1290 - \left(1650 \times \tan 10°\right)}$$

$$\mu = \frac{1877.46}{999.06}$$

$$\mu = 1.879$$

2. In an orthogonal cutting operation on a work piece of width 2.5mm, the uncut chip thickness is 0.25mm and the tool rake angle is zero degree, it was observed that the chip thickness was 1.25mm. The cutting force was measured to be 900N and the thrust force is 810N. Let us calculate the shear strength of the work piece material. If the coefficient of friction between the chip and the tool is 0.5. Let us also determine the machining constant C_m.

Solution:

Given Data:

- $b = 2.5 \text{ mm}$
- $T_1 = 0.25 \text{ mm}$
- $\alpha = 00$
- $T_2 = 1.25 \text{ mm}$
- $F_z = 900/V$
- $F_x = 810/V$
- $\mu = 0.5$

To find:

- Shear Strength

- Machining constant

Chip thickness ratio,

$$r = \frac{t_1}{t_2} = \frac{0.25}{1.25} = 0.2$$

Shear angle,

$$\beta = \tan^{-1}\left[\frac{r\cos\alpha}{1 - r\sin\alpha}\right]$$

$$= \tan^{-1}\left[\frac{0.2\cos 0}{1 - 0.2\sin 0}\right]$$

$$\beta = 11.31°$$

Shear force,

$$F_o = F_z \cos\beta - F_x \sin\beta$$

$$= 900\cos 11.31 - 810\sin 11.31$$

$$F = 723.66/N$$

Shear Stress or strength,

$$\tau_o = \frac{F_o}{A_z} = \frac{723.66}{2.5 \times 0.25} \times \sin 11.31 \quad [\because A_1 = 0.25 \times 2.5]$$

$$\tau = 227 N/mm^2$$
$$\mu = \tan^{-1}(A_1)$$
$$\gamma = \tan^{-1}(0.5) = 26.56$$

Machining constant,

$$C_m = 2 \times 11.31 + 26.56 - 0$$

$$C_m = 49.18°$$

Result:

- Shearing Strength $\beta = 227 N/mm$.
- Machining Constant $C_m = 49.18°$.

1.4   Tool Wear and Tool Failure

Tool Failure

A cutting tool is said to have failed when it ceases to function satisfactorily. Tool failure may be classified as:

- Catastrophic failure.

- Gradual or progressive wear.

Catastrophic failure of tool will occur when the cutting force acting on the tool exceeds the critical strength of the tool material and tool fails without giving any indication.

Under normal cutting conditions, the tool is subjected to gradual or progressive wear. As soon as the cutting operation is started, gradual or progressive wear also starts and it progresses with the machining process. The progressive wear may happen because of crater formation leading to crater wear or flank wear.

Classification of Tool wear

The tool wear is generally classified as follows:

- Flank wear or crater wear

- Face wear

- Nose wear

Mechanisms of Tool Wear

Attrition Tool Wear

When a softer metal slide over a harder metal such that it always presents a new-ly-formed surface to the same potential of the hard metal . On account of friction, high temperature and pressure, particles of the softer material surface adhere to a few high spots of the harder metal.

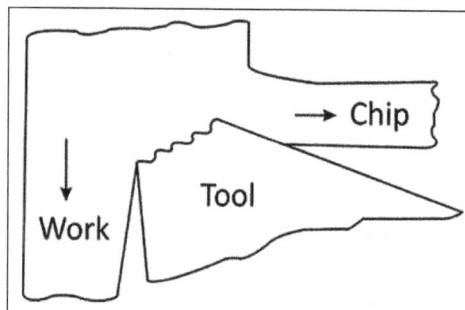

Attrition tool wear.

As a result, flow of the softer metal over the surface of the hard metal becomes irregular or less laminar and contact between the two becomes less continuous. More particles join up with those already adhering and a so called built-up edge is formed. Sooner or later some of these fragments which may have grown up to microscopic size are torn from the surface of the hard metal. When this process continues for some time, it appears as if the surface of the hard metal has been nibbed away and made uneven.

Diffusion Tool Wear

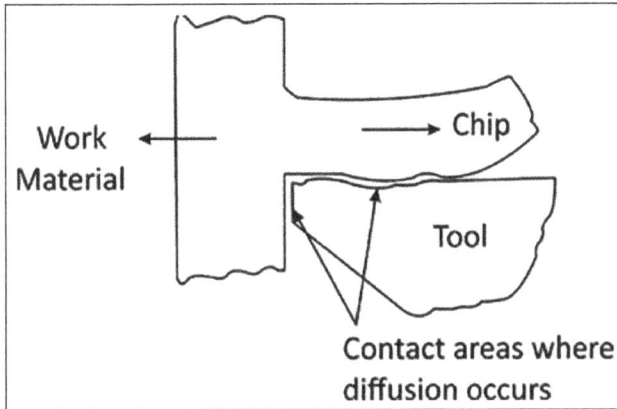

Diffusion tool wear.

When a metal is in sliding contact with another metal and the temperature at their interface is high, conditions may become right for atoms from the harder metal to diffuse into the softer matrix, thereby increasing the latter's hardness and abrasiveness.

On the contrary atoms from the softer metal may also diffuse into the harder medium weakening the surface layer of the latter to such an extent that particles on it are dislodged, torn (or sheared off) and are carried away by the flowing medium.

Fatigue Wear

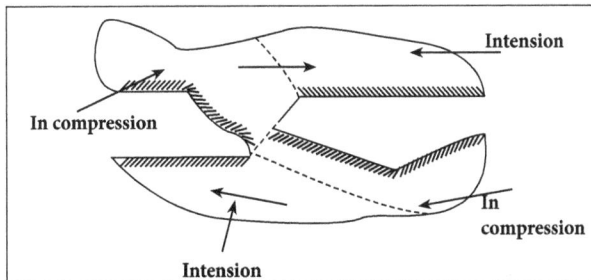

Fatigue wear.

When two surfaces slide in contact with each other under pressure, asperities on one surface interlock with those of the other. Due to the frictional stress, compressive stress is produced on one side of each interlocking asperity and tensile stress on the other side. After a given pair of asperities have moved over or through each other, the above stresses are relieved. New pairs of asperities are, however, soon formed and the stress cycle is repeated. Thus the material of the hard metal near the surface undergoes cyclic stress.

This phenomenon causes surface cracks which ultimately combine with one another and lead to the crumbling of the hard metal. Further, the hard metal may also be subjected to variable thermal stress owing to temperature changes brought about by cutting fluid, chip breakage and variable dimensions of cut, again contributing to fatigue wear.

1.4.1 Tool Life

Tool life is defined as the time for which a tool can cut effectively or it is the time between two successive re-sharpening of a cutting tool. Cutting speed has more influence on tool life when compared to feed and depth of cut. The tool life as a function of cutting speed is shown in the figure below.

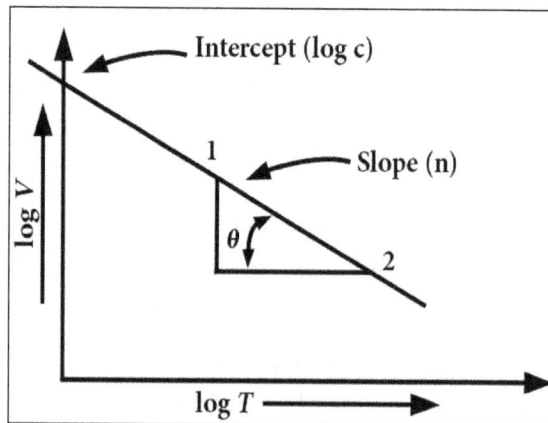

Tool life graph.

The logarithm of tool life in minutes when plotted against the logarithm of cutting speed in m/min, the resulting curve is very nearly a straight line in most instances. For all practical purposes it can be considered as a straight line.

Parameters that Influence the Life of Tool

Tool life is the time a tool will operate satisfactorily until it is dulled. If a cutting tool is to have a long life, it is estimated that the face of the tool be as smooth as possible.

The life of the cutting tool is affected by the following factors:

1. Cutting speed: Cutting speed has the greatest influence on tool life. As the cutting speed increases the temperature also raises, the heat is more concentrated on the tool than on the work and also increase in the hardness of the work accelerate the abrasive action.

$$V \, T^n = C$$

Or,

$$n = \tan \theta = \frac{\log v_1 - \log v_2}{\log T_2 - \log T_2}$$

Where,

V = Cutting speed

T = Tool life in min

n = Exponent

C = Constant

2. Feed and depth of cut: The tool life is influenced by the feed rate also with a fine feed the area of chip passing over the tool face is greater than that of a coarse feed for a given volume.

3. Tool geometry: The tool life is also affected by tool geometry, tool with large rake angle becomes weak as a large rake reduces the tool cross-section and the amount of metal to absorb the heat. The nose radius tends to improve tool life. The effect of clearance is to improve tool life. The optimum clearance is between 10' to 15'.

4. Tool material: The effect of tool material on life indicates that higher cutting speed is not only criteria considered for removing large volume of metal. What is desirable is the high rate at which the stock will be removed per cutting edge or tool life.

5. Cutting fluids: Cutting fluids affect tool life to a greater extent. A cutting fluids does not only carry away the heat generated and keep the tool, chip and work piece cool, but reduces the coefficient of friction at the chip tool interface and increase tool life.

1.4.2 Effects of Cutting Parameters on Tool Life

The metal removal in any machining operation is equal to the product on the operating parameters, namely, cutting speed, v_c, feed rate, f and depth of-cut, b. Hence an equal change in any of the three operating conditions will equally affect the metal removal and the tool life as well. Increase in any of the operating parameters in any degree will reduce tool life and the extent of loss in tool life depends on which specific parameter has been increased.

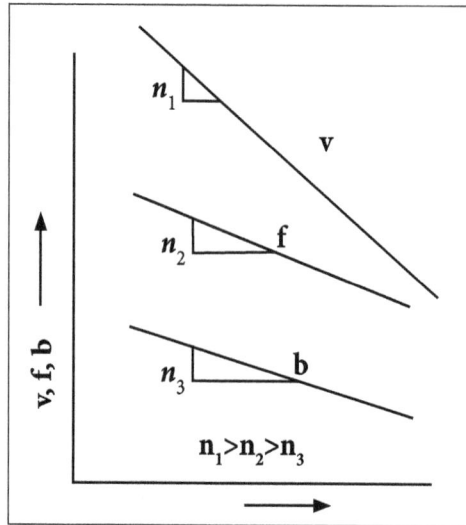

Effects of cutting parameters on tool life.

The figure above, a depicts the effect of the parameters on tool life. The slope of the curve is highest for cutting speed, that is for feed rate intermediate and that for depth-of-cut the least. That is, in general, tool life is less affected by changes in depth-of-cut than by changes in either of the other two operating conditions and that the effect of changes in feed rate is greater than that for change in depth-of-cut. Finally, cutting speed/velocity has a greater effect on tool life than either depth-of-cut or feed rate.

To express the above in a quantified manner, the comparative significance is as follows:

For a 50 percent increase in depth-of-cut, there is only a 15 percent reduction in tool life. If the feed rate is increased by 50 percent, tool life will decrease approximately by 60percent and a 50 percent increase in cutting speed will generally reduce tool life by about 90 percent.

Hence, as a general strategy to obtain the best compromise between tool life and metal-removal rate, depth-of-cut should be maximum. Similarly, for optimizing trade-off between machining time and tool life, feed rate should be at the highest level possible. Finally, the determination of cutting velocity is most critical for establishing operating conditions.

However, there are certain limitations. The surface finish requirement may limit the feed rate or the strength of the insert to withstand cutting forces may also limit feed rate. Similarly, the capability of the machine tool, part, cutting tool and framing to resist cutting forces may limit feed rate and depth-of-cut. Again, the amount of stock that needs to be removed from the part may limit depth-of-cut.

The cutting force is dependent on several factors besides the feed rate, Unit force on the cutting edge is best indicated by the maximum unreformed. chip thickness. Increase in feed rate will increase, unreformed chip thickness, which is also affected by changes in lead angle. If lead angle is increased, feed can be increased, which is limited by the strength of the cutting edge.

Additionally, available machine horsepower often limits the metal removal rate. On limited horsepower applications, depth-of-cut and feed rate should be maximized and the cutting speed so set that the horsepower drawn is, within limits. However, the cutting speed should never be set so low that built-up edge becomes a problem.

Ways of Expressing Tool Life

- Volume of metal removed per grind.

- Number of work pieces machined per grind.

- Time unit.

- Rigidity of work, tool and machine.

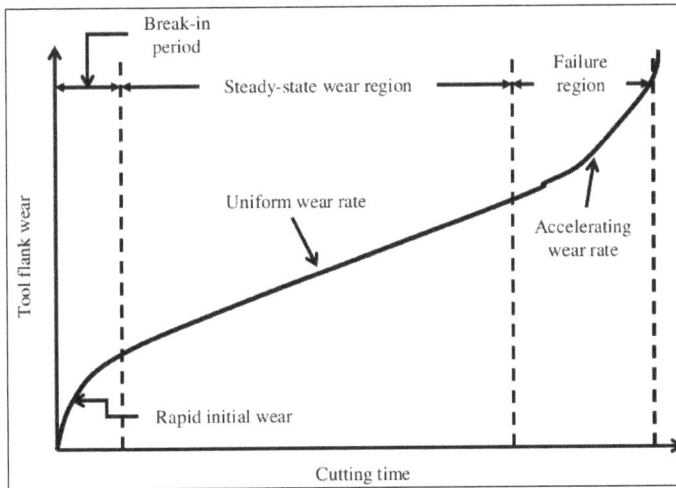

Reasons for Flank Wear in Cutting Tool

- Flank wear takes place when machining brittle materials like Combustion engine.

- Flank wear takes place when feed is less than 0.15 mm/sec.

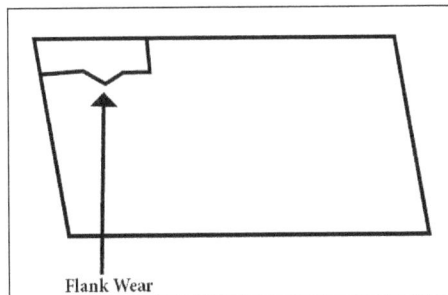

Flank wear.

Built Up Edge

During cutting, the temperature and pressure is quite high, it causes the chip material to weld itself to the tool face and cutting tip. This welded portion is called as Built-up-Edge.

1.4.3 Tool Failure Criteria and Taylor's Tool Life equation

Tool Failure Criteria

The following are some of the possible tool failure criteria that could be used for limiting tool life:

Based on tool wear:

- Wear land size.

- Crater depth, width or other parameters.

- A combination of the above two.

- Chipping or fine cracks developing at the cutting edge.

- Volume or weight of material worn off the tool.

- Total destruction of the tool.

Based on consequences of worn tool:

- Limiting value of change in component size.

- Limiting value of surface finish.

- Fixed increase in cutting force or power required to perform a cut.

The actual tool life values obtained would depend on the failure criterion adopted. Typical wear land sizes used as tool life limits are shown in the table.

Table: The various possible tool life criteria.

Wear land, mm	Tool material	Remark
0.75	Carbides	Roughing
0.25 to 0.38	Carbides	Finishing
1.50	HSS	Roughing
0.25 to 0.38	HSS	Finishing
0.25 to 0.38	Oxide	Roughing and Finishing

The main advantage of wear land as a failure criterion is that it is fairly easy to measure and is directly proportional to the surface finish and cutting forces. But practically it is not a uniform one and hence is at its maximum. At high speeds and feeds, diffusion wear will be more and as such the crater wear also may provide a failure criterion.

But the problem is with the measurement. The crater is to be scanned for maximum depth which is laborious and time consuming. The total destruction criterion is easier to apply but then there are associated problems, e.g. in carbides it may spoil the nearby edges and also do damage to the work . There-fore a large wear land size can be equivalent to total destruction.

Alternatively indirect methods such as the fixed increase in the value of cutting force. Or increase in the power consumed or measured vibrations, etc. Can be used as an indication of the end of tool life. These are more generally used in CNC machine tools where automatic tool life monitoring facilities are present.

Taylor's Tool Life Equation

Taylor thought that there is an optimum cutting speed for best productivity. This he reasoned from the fact that at low cutting speeds. The tools have higher life but productivity is low and at higher speeds the reverse is true.

This inspired him to check up the relationship of tool life and cutting speed. Based on his experimental work he proposed the formula for tool life. The relation between cutting speed and tool life is expressed by Taylor's formula which states:

$$V T^n = C... \text{ i} \qquad\qquad ...(i)$$

Where,

V = Cutting speed

T = Tool life in min

n = Exponent

C = Constant

Equation (i) can be rewritten as,

$$\log r + n \log T = \log c \qquad\qquad ...(ii)$$

$$\text{or} \qquad \log T = \left(\frac{1}{n}\right)\log c - \left(\frac{1}{n}\right)\log r$$

As a log-log graph: The Taylor's tool life equation represents a straight line. Equation (i) can be generated to include the effects of feed and depth of cut. One such relationship is of the form.

$$VT^n \ f^{n1} \cdot d^{n2} = C_1 \qquad \qquad \text{...(iii)}$$

Where, the exponents n, n1 and n2 and the constant C_1 depend upon tool and work material, tool geometry and type of coolant.

Problems

1. A cutting tool when used for machining work piece at a cutting speed of 50 m/min lasted for 100 minutes. Taking n = 0.26 in the Taylor's tool-life equation, let us determine (1) the life of the tool for an increase in cutting speed by 25% and (2) the cutting speed to obtain a tool life of 180 minutes.

Solution:

Given data:

Cutting speed V = 50 m/min

Time T = 100 min

n = 0.26

Taylor's equation $V \ T^n = C$

$50 \times 100^{026} = C$

C = 165.57

Life of the tool for an increase in cutting speed by 25%.

Now,

$$V = 50 + \left(50 \times \frac{25}{100} \right)$$

$$V = \ 62.5 \ m / min$$

$$\Rightarrow 62.5 \times T^n = 165.57$$

$$T^n = \ 2.649$$

$$T = \ 2.649)^{1/0.26}$$

$$T = \left(2.649 \right)^{3846}$$

$$T = \ 42.38 \ min$$

The cutting speed to obtains tool life of 180 minutes:

$$V T^n = C$$

$$V \times 180^{0.26} = 165.54$$

$$V = \frac{165.57}{3.858}$$

$$V = 42.9 \text{ m} / \text{min}$$

2. The Taylor Ian tool-life equation for machining C-40 steel with a 18:4:1 H.S.S. cutting tool at a feed of 0.2 mm/min and a depth of cut of 2 mm is given by $VT^n=C$, where n and C are constants. The following V and T observations have been noted.

V_1 m/min	25	35
T_1 min	90	20

Let us calculate:

- n and C.

- Let us also find the cutting speed for a desired tool life of 60 minutes.

Solution:

Given Data:

Cutting Tool = C-40 Steel with 18:4:1 HSS

Feed Rate = 0.2 mm/min

Depth of cut = 2 mm

Cutting Speed = 25 and 35 m/min

Tool Life = 90 and 20 min

The given Taylorian tool-life equation is $V T^n = C$

(i) To find n and c:

$$V_1 T_1^n = C$$
$$V_2 T_2^n = C$$

From the given data,

$$25 \times 90^n = C \qquad\qquad \text{...(1)}$$

$$35 \times 20^n = C \qquad\qquad \text{...(2)}$$

From Equation (1) and (2),

$$\left(\frac{90}{20}\right)^n = \frac{35}{25}$$

$$25 \times 90^n = 35 \times 20^n$$

$$4.5^n = 1.4^n$$

$$n = \log 4.5 = \log 1.4$$

$$n = 0.224$$

In equation (1)

Substituting the value of n,

$$25 \times 90^{0.224} = C$$

$$\therefore C = 68.5$$

(ii) Recommended cutting speed for 60 min tool life:

Tool Life Equation $V T^n = C$

$$V\, 60^{0.224} = 68.5$$
$$V \times 2.502 = 68.5$$
$$V = 27.38 \text{ m / min}$$

Result:

(i) Value of n = 0.224

$$c = 68.5$$

(ii) For 60 min tool life,

Cutting speed V = 27.38 m/min

3. During straight turning of a 24 mm diameter steel bar at 300 r.p.m. with an H.S.S. tool, a tool life of 9 min. was obtained. When the same bar was turned at 250 r.p.m., the tool life increased to 48.5 min. Let us determine the tool life at a speed of 280 r.p.m.

Solution:

Given Data:

Diameter D- 24 mm

Speed N_1 - 300 rpm

Tool Life $T_1 = 9$ min

Speed $N_2 = 250$ rpm

Tool life $T_2 = 48.5$ min

To find:

Tool life at a speed of 280 rpm,

$$V_1 = \frac{\pi D N_1}{1000} = \frac{\pi \times 24 \times 300}{1000} = 22.62 \, \text{m} / \text{min}$$

$$V_2 = \frac{\pi D N_2}{1000} = \frac{\pi \times 24 \times 250}{1000} = 18.85 \, \text{m} / \text{min}$$

$$V_3 = \frac{\pi D N_3}{1000} = \frac{\pi \times 24 \times 280}{1000} = 21.11 \, \text{m} / \text{min}$$

Tool Life of 280 rpm

By using Taylor's Tool Life equation,

$$V_1 T_1^n = V_2 T_2^n$$

$$\frac{T_1^n}{T_2^n} = \frac{V_2}{V_1}$$

$$n \, \text{In} \left(\frac{T_1}{T_2} \right) = \text{In} \left(\frac{V_2}{V_1} \right)$$

$$n = \frac{\text{In} \left(\dfrac{V_2}{V_1} \right)}{\text{In} \left(\dfrac{T_1}{T_2} \right)}$$

$$n = \frac{\text{In} \left(\dfrac{18.85}{22.62} \right)}{\text{In} \left(\dfrac{9}{48.5} \right)}$$

$$n = 0.108$$

Now,

$$V_1 T_1^n = V_3 T_3^n$$

$$T_3 = \left(\frac{V_1}{V_3} \right)^{1/n} \times T_1.$$

Cutting Tool Materials

2.1 Desired Properties and Types of Cutting Tool Materials

Various tool materials used in industry are:

- Carbon Steel
- Cemented Carbides
- High Speed Steel
- Ceramics
- Abrasives
- Diamond

A cutting tool must have the following characteristics in order to produce good quality and economical parts.

Essential Requirements of a Tool Material

- Hot hardness: It is the ability to retain its hardness at high temperatures. The tool must maintain its hardness at high temperature.

- Toughness: The tool material should have sufficient toughness to withstand shock and vibrations. This property limits the hardness of the tool.

- Wear resistance: It is the ability to resist wear. Addition of cobalt increases the wear resistance property of the tool.

- Low friction: The co-efficient of friction between tool and the work piece must be less. This will reduce friction and tool wear.

- Cost of tool material: It should be easy to manufacture the tool from the material.

High Speed Tool Steel

The need for tool materials to withstand increased cutting speeds and temperatures led to the development of high-speed tool steels (HSS). The major difference between HSS and plain high carbon steel is the addition of alloying elements to harden and strengthen the steel and make it more resistant to heat.

The most generally used alloying elements are manganese, chromium, tungsten, vanadium, molybdenum, cobalt and niobium. While each of these elements will add certain specific desirable characteristics, it can be generally stated that they add deep hardening capability, high hot hardness, resistance to abrasive wear and strength, to HSS. These characteristics allow relatively higher machining speeds and improved performance over plain high carbon steel.

Cast Alloys

The alloying elements in HSS principally cobalt, chromium and tungsten improve the cutting properties sufficiently, that metallurgical researchers developed the cast alloys, a family of materials without iron.

A typical composition for this class was 45% cobalt, 32% chromium, 21% tungsten and 2% carbon. The purpose was to obtain a cutting tool with hot hardness superior to HSS.

When applying cast alloy tools, their brittleness should be kept in mind and sufficient support should be provided at all times. Cast alloys provide high abrasion resistance and are thus useful for cutting scaly materials or those with hard inclusions.

Manufacture of Carbide Products

The term "tungsten carbide" describes a comprehensive family of hard carbide compositions used for metal cutting tools, dies of various kinds and wear of parts. In general, these materials are composed of the carbides of tungsten, titanium, tantalum or some combination of these, sintered or cemented in a matrix binder, usually cobalt.

Classification of Carbide Tools

Cemented carbide products are classified into three major categories:

- Wear Grades: It is used primarily in dies, machine and tool guides and in everyday items such as line guides on fishing rods and reels. Used anywhere, good wear resistance is required.

- Impact Grades: It is also used for dies, particularly for stamping and forming and in tools such as mining drill heads.

- Cutting Tool Grades: The cutting tool grades of cemented carbides are divided into two groups, depending on their primary application. If the carbide is intended for use on cast iron that is a non ductile material, it is graded as cast iron carbide. If it is to be used to cut steel, a ductile material, it is graded as steel grade carbide.

Cast iron carbides must be more resistant to abrasive wear. Steel carbides require more resistance to cratering and heat. The tool wear characteristics of various metals are different, thereby requiring different tool properties.

Abrasives

Abrasive grains such as aluminum oxide or silicon carbide are used in the grinding wheels. They are used to remove a very small portion in the work piece and usually grinding is employed as a final operation.

The above said abrasives are called manufactured abrasives. There are natural abrasives such as emery, corundum diamond and quartz.

Optimum Temperature of Each Tool Materials

- Carbon Steel - 200°C - 250°C.
- HSS - 600°C - 620°C.
- Converted Carbide - 1000°C.
- Ceramics - Up to 1200°C.
- Diamond -Up to 1650°C.

Advantages of Diamond Tools

- Cutting speed is 50 times higher than HSS.
- It can resist temperature up to 1250°C.
- It conducts heat quickly.
- It has low co-efficient of friction.

Different Cutting Tool Materials Used in Metal Cutting

1. High Carbon Steel:

Plain carbon steel having a carbon percentage up to 1.5% are not suitable for tools used in production work. It is mainly used for hand tools. It is less costly, easily forgeable and easy to heat treat. It lose hardness at 300°C and they are not suitable for high speed cutting and heavy duty work.

2. High Speed Steel:

It is a special alloy steel which increases strength, toughness, wear resistance, cutting ability, for alloying purpose, materials like tungsten, chromium, vanadium, cobalt, molybdenum are used.

High speed steel tools can operate at 2 to 3 times higher speeds than high carbon steel tools. These tools retains cutting ability at operating temperatures of 600°C.

3. Cemented Carbides:

These are used for mass production. These carbide tools are formed by tungsten, titanium or tantalum with carbon and the compound is combined with cobalt and sintered in furnace at 1400°C.

It possess a very high degree of hardness and wear resistance. It can be operated at speeds 5 to 6 times higher than high speed steels.

4. Stellite:

It is a non-ferrous alloy consisting of cobalt, tungsten and chromium. It has good shock resistance. Tools made by stellite can operate at two times faster than the high speed steel tools.

5. Cemented Oxides or Ceramics:

Tools made of ceramic material are capable of withstanding high temperatures, without losing hardness. They are more wear resistant than cemented carbide tools. They are not used for rough machining work because they are more brittle and have low bending resistance.

While machining with ceramic tools, coolant is not required. These tools are made from sintered aluminum oxide and various boron-nit-ride powders. These powders are mixed together and sintered at 1700°C.

6. Diamond:

It is the hardest cutting tool material. It offers high wear resistance but low shock resistance due to brittleness. As diamonds have low coefficient of friction, they are used for high grade super finishing.

It offers the highest tool life, 50-100 times more than cemented carbides. The main disadvantages of diamonds are their brittleness and high cost.

Advantages of Cemented Carbides Tools

- High heat and wear resistance.

- Cemented carbide tipped tools can machine metals even when their cutting elements are heated to a temperature of 1000°C.

- It can withstand cutting speed 6% or more than '6' times higher manufactured material and has extremely high compressive strength.

Limitation of Cemented Carbide Tool

- It is brittle material.

- Low resistance to shock.

- It must be very rigidly supported to prevent cracking.

Cubic Boron Nit-ride

Advantages:

- High hardness.

- High thermal conductivity.

- High tensile strength.

2.2 Cutting Fluids: Desired Properties

A cutting fluid may be defined as any substance (may be liquid, gas or solid) which is applied to a tool during a cutting operation to facilitate removal of chips.

Requirements of a Cutting Fluid

A cutting fluid should possess the following qualities in order to be of practical value:

- It should have long life, free of excessive oxide formation that might clog circulation system.

- It should be suitable for a variety of cutting tools and materials and the cutting operations.

- It should have lubricating qualities, high thermal conductivity and low viscosity to permit easy flow and easy separation from impurities and chips and should not stick to work piece or machine.

- It should be transparent where high dimensional accuracy and fine finish are required in order to enable the operator to have a clear view of tool and work piece.

- It should present no fire or accident hazards or emit abnoxious odours or vapours harmful to operator or work piece and should cause no skin irritation.

Desired Properties

Properties and Purpose of Cutting Fluids

The following are the essential purposes of cutting fluids:

- To reduce friction: The cutting fluid reduces the friction at the tool chip interface and also at the tool work interface. Since the coefficient of friction is reduced at the tool chip interface the chip flow is increased. If on the other hand, if the coefficient of friction is more not only the chip finds difficult to flow, but also it increases the power consumption.

- To improve surface finish: Since there is a smooth flow of chip and lesser coefficient of friction, surface finish is improved.

- To cool the tool and the work piece: The cutting action produces more heat to be generated at the zone of chip, tool and work piece. The cutting fluid reduces the heat. Since heat is reduced, it enables the hot hardness of the tool to be retained and at the same tin tool life is also increased.

- To move the chips quickly: The flow of fluid enables the quick disposal of the chips. For example, in drilling, the chip flows through the flutes.

Important Properties of Fluids

- It should absorb more heat.
- It should reduce friction.
- It should not be corrosive in nature.
- It should have low viscosity.
- It should be economical.

Various methods to be applied while using the cutting fluids:

1. Manual Application

Application of a fluid from a can manually by the operator. It is not acceptable even in job-shop situations except for tapping and some other operations where cutting speeds are very low and friction is a problem. In this case, cutting fluids are used as lubricants.

2. Flooding

In flooding, a steady stream of fluid is directed at the chip or tool-work piece interface. Most machine tools are equipped with a recirculating system that incorporates filters for cleaning of cutting fluids. Cutting fluids are applied to the chip although better cooling is obtained by applying it to the flank face under pressure.

The left image shows the Rake face flooding by means of a recirculating system and right image shows the Rank face application of the cutting fluid.

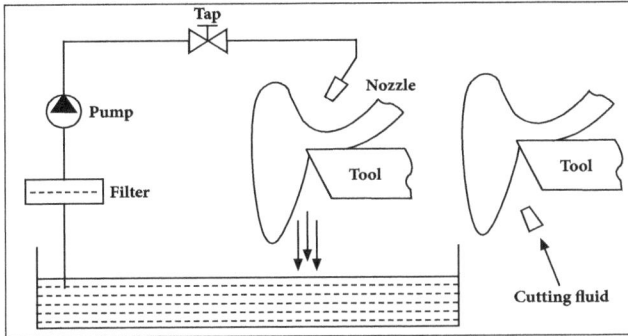

Cutting fluid application.

3. Coolant-fed Tooling

Some tools, especially drills for deep drilling, are provided with axial holes through the body of the tool so that the cutting fluid can be pumped directly to the tool cutting edge.

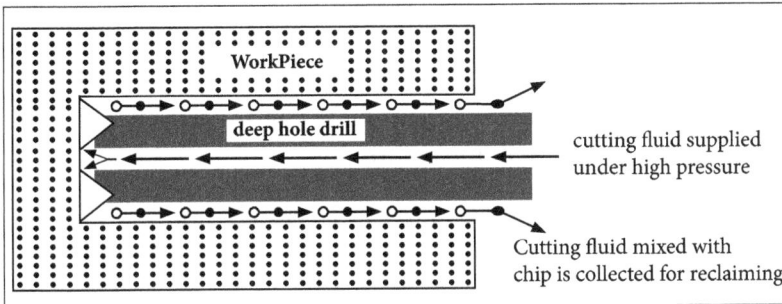

Internal cutting fluid application in deep hole drilling.

4. Mist Applications

Fluid droplets suspended in air provide effective cooling by evaporation of the fluid. Mist application in general is not as effective as flooding, but can deliver cutting fluid to inaccessible areas that cannot be reached by conventional flooding.

Main Functions of Cutting Fluids

- To cool the tool: Cooling the tool is necessary to prevent metallurgical damage and to assist in decreasing friction at the tool-chip interface and at the tool-work piece interface.

- To cool the work piece: The role of the cutting fluid in cooling the work piece is to prevent its excessive thermal distortion.

- To lubricate and reduce friction: The energy or power consumption in removing metal is reduced. Abrasion or wear on the cutting tool is reduced, thereby increasing the life of the tool. By virtue of lubrication, less heat is generated and the tool, operates at lower temperature.

- To improve surface finish.

- To protect the finished surface from corrosion.

- To cause chips break up into small parts rather than remain as long ribbons which are hot and sharp and difficult to remove from the work piece.

- To wash the chip away from the tool.

2.2.1 Types and Selection

The broad classification of cutting and grinding fluids is as follows:

- Straight or neat-oils.

- Water-miscible cutting fluids.

- Synthetics or semi-chemical cutting fluids.

1. Straight or Neat-oils:

- These are derived from petroleum, animal, marine or vegetable substances and may be used straight or in combination.

- Their main function is lubrication and rust prevention.

- They are chemically stable and lower in cost.

- They are usually restricted to light-duty machining on metals of high machinability, such as aluminum, magnesium, brass and leaded steels.

2. Water Miscible Cutting Fluids:

Water miscible fluids form mixtures ranging from emulsions to solutions, which due to their high specific heat, high thermal conductivity and high heat of vaporization, are used on about 90% of all metal-cutting and grinding operations:

- The water blend is usually in the ratio of one part oil to 15 to 20 parts water for cutting and 40 to 60 parts water for grinding.

3. Synthetics or Semi-chemical Cutting Fluids:

- These can give high metal removal rates and more life than other machining fluids.

- These are all cannot be reversed and also they are less costly.

- It is mostly used in grinding operations.

Selection of Cutting Fluids

The selection of cutting fluid for a given application requires the examination of a number of parameters such as:

- Workplace material.

- Machining operation.

- Cutting tool material.

- Other ancillary factors.

Table: Selection of cutting fluid based on work material.

Materials	Characteristics	Cutting fluid to be used
Grey cost iron	Grey cast iron could be machined dry since the graphite flakes act as solid lubricants in cutting.	Soluble oils and thinner neat oils am satisfactory for flushing swarf and metal dust.
Copper alloy	Better machinability. Some could be machined dry.	Water-based fluids could be conveniently used. For tougher alloys, a neat oil blended with fatty or inactive EP additive is used.
Aluminium alloy	They are generally ductile and can be machined dry. However, in combination with steel they have a high friction coefficient. Generally, BUE forms on the tool and prevents the chip flowing smoothly away from the work.	It is necessary to have the tool surface highly polished. Generally, a soluble oil to which an oiliness agent has been added is used. For more difficult application, straight neat oil or fatty oil or kerosene could be used.
Mild steel and low to medium -carbon steels	Easiest to machine. Low carbon steels have lower tensile strength and hence, may create problems because of their easy tearing.	Milky-type soluble oil or mild EP neat cutting oil could be used.
High carbon steels	They are tough and pressures and temperatures in the cutting zone are higher.	EP cutting oil is used. In some applications, milky soluble oils can be used.
Alloy steels	Particularly with chromium and nickel, the steels are tough. They are similar to high-carbon steels.	EP cutting oil is used. In some applications, milky soluble oils can be used.
Strain less steel and heat resistant alloys	They have great toughness, corrosion resistance and pronounced work hardening. Very difficult to machine. Because of their toughness, they create very high pressure in the work area.	Use high-performance neat oils with high concentration of chlorinated additives. Sulphur additives are to be avoided. Some high-performance EP soluble oils may sometimes be useful.

Table: Selection of cutting fluid based on tool material.

Tool Material	Characteristics	Cutting Fluid Requirement
High-carbon steel	Not widely used. They should be well cooled.	Water-based coolants are generally used.
High-speed steels	Have better hot hardness characteristics.	For general machining water-based cutting fluids can be used. For heavy-duty work, El' neat oils are preferable.
Non-ferrous materials	During a cut they should never be overcooled or subjected to intermittent cooling because they are brittle and are likely to suffer thermal shock.	Neat cutting oils are the most suitable choice for most applications.
Carbides, ceramics and diamond	These are used for very high speeds. Hence the requirement is to reduce the large amount of heat produced to reduce the thermal distortion of the work piece.	Water-based coolant is recommended. In low-speed applications, EP based oils could reduce the problem of adhesion of chip with tool.

2.3　Heat Generation in Metal Cutting

The heat generated during cutting operation depends on the rate of metal cutting, cutting speed, specific heat and thermal conductivity of the work piece and tool material. Amongst all these parameters mentioned, cutting speed has more influence on the temperature rise during operation hence heat generated.

This is because as speed increases, the time for heat dissipation decreases, thus temperature increases. The total heat generated Q during a machining operation is distributed between work piece, tool, chips and surrounding.

$$Q_{total} = Q_w, Q_T, Q_C \text{ and } Q_S$$

Where, Q_w, Q_T, Q_C and Q_S is the amount of heat conducted into the work piece, tool, chips and is the amount of heat dissipated to the surroundings respectively. The amount of heat dissipated to the surroundings is very small and can be neglected.

The heat distributed in metal cutting operation is approximately 80 : 10 : 10 between chips, tool and work piece. This ratio shows that the majority of heat generated during the process is carried away by the chips. During machining, heat is generated at the cutting point from three sources, as indicated in figure (a). Those sources and causes of development of cutting temperature are:

- Primary shear zone (1), where the major part of the energy is converted into heat.

- Secondary deformation zone (2) at the chip – tool interface where further heat is generated due to rubbing or shear.

- At the worn outflanks (3) due to rubbing between the tool and the finished surfaces.

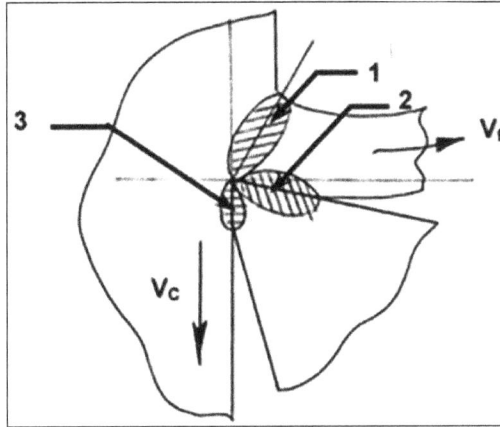

(a) Sources of heat generation in machining.

The heat generated is shared by the chip, cutting tool and the blank. The apportionment of sharing that heat depends upon the configuration, size and thermal conductivity of the tool work material and the cutting condition. Figure (b) shows that maximum amount of heat carried away by the flowing chip.

Around 10 to 20% of the total heat goes into the tool and some heat is absorbed in the blank. With the increase in cutting velocity, the chip shared heat increases.

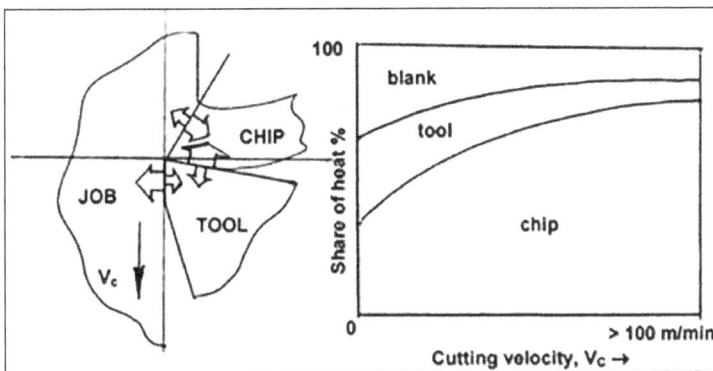

(b) Apportionment of heat amongst chip, tool and blank.

2.3.1 Factors Affecting Heat Generation

All the mechanical work done in cutting metal is converted into an equivalent amount of heat. The work (W) done in cutting depends upon the cutting Force (P_2) and cutting speed (V) it is determined from the formula,

$$W = P_2 \, V kgf / min$$

The amount of heat generated in a unit of time by metal cutting is based upon the thermal equivalent of work equal to 427 kgf m per k cal,

$$Q = \frac{W}{427} = \frac{P_2 \cdot V}{427} \text{ kcal per min.}$$

The generated heat is distributed between the workers chip and tool, only negligible amount of heat is dissipated to the ambient air.

The main sources of heat in metal cutting are:

- The shear zone.

- The chip-tool interface zone.

- The work tool interface zone.

(1) The Shear Zone:

Where the main plastic deformation takes place due to shear energy. This heat raises the temperature of the chip. Part of this heat is carried away from the chip when it moves upward along the tools.

Considering a continuous type chip, as the cutting speed increases for a given rate of feed, the chip thickness decreases and less shear energy are required for chip deformation. So the chip is heated less from this deformation.

(2) The Chip-Tool Interface Zone:

Where secondary plastic deformation due to friction between heated chip and tool takes place. This causes a further rise in the temperature of the chip. The frictional heat increases with increasing cutting speed. The tool-chip interface temperature increases with the cutting speed and the work hardness because the heat is concentrated upon a smaller area even through the quality of heat transferred to be remained constant.

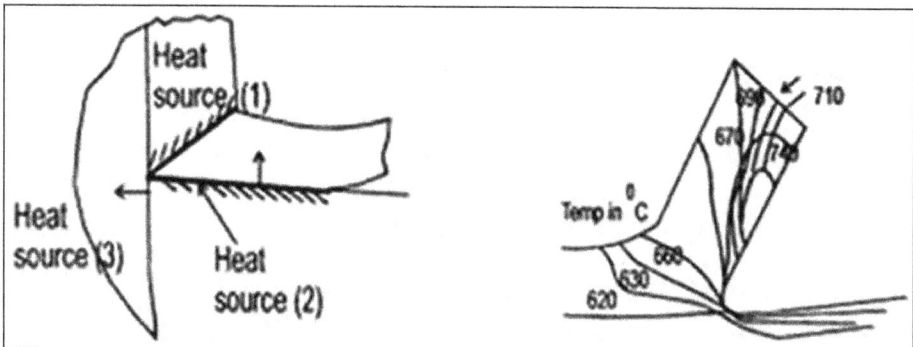

Source of heat in metal cutting Temperature distribution in cutting zone.

The tool chip temperature rise but less rapidly than for a rise in the cutting speed.

Changes in the depth of cut is appreciably greater than the nose radius. Legs heat is generated when higher feed rates are used but the surface quality is adversely affected.

(3) The Work-Tool Interface Zone:

At flanks where friction rubbing occurs,

From the figure,

TC = Total heat in tool and chip,

C = Heat in chips,

WTC = 100% total heat.

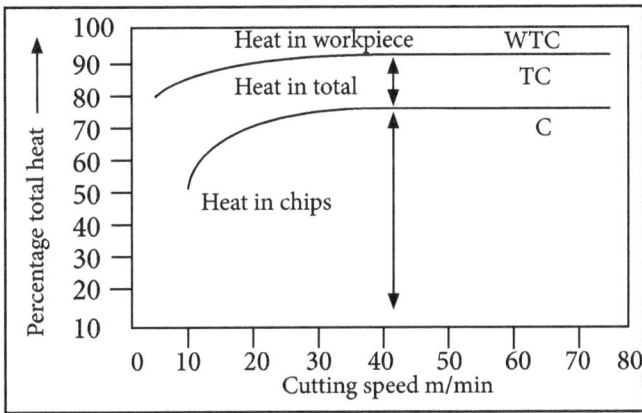

The above figure shows temperature distribution during the cutting operation. In machining steel with a single point tool having a cemented carbide tip, the relative amount of heat passing in to the chip, work piece and tool at different cutting speeds is in the above figure.

It is seen that as the cutting speed increases, proportionally more heat is carried away by the chip and less is transferred to the work piece and the tool. The fact that at high cutting speeds most of the heat energy is carried away by the chip leads to the advantage and practicability of the tool is tolerable.

2.3.2 Heat Distribution in Tool and Work Piece and Chip

Heat is generated at tool-chip interface by friction. The intensity, IC of the frictional heat source is approximately given by,

$$I_c = FV_x/hb$$

Where,

F = Friction force,

V_x = Sliding velocity of the chip along the interface,

h = Plastic contact length.

The effect of the cutting temperature, particularly when it is high, is mostly detrimental to both the tool and the job. The major portion of the heat is taken away by the chips. But it does not matter because chips are thrown out.

So attempts should be made such that the chips take away more and more amount of heat leaving small amount of heat to harm the tool and the job. The possible detrimental effects of the high cutting temperature on cutting tool (edge) are:

- Rapid tool wear, which reduces tool life.

- Plastic deformation of the cutting edges if the tool material is not enough hot-hard and hot-strong.

- Thermal flaking and fracturing of the cutting edges due to thermal shocks.

- Built-up-edge formation.

The possible detrimental effects of cutting temperature on the machined job are:

- A dimensional inaccuracy of the job due to thermal distortion and expansion-contraction during and after machining.

- A surface damage by oxidation, rapid corrosion, burning etc.

- Induction of tensile residual stresses and microcracks at the surface/subsurface.

However, often the high cutting temperature helps in reducing the magnitude of the cutting forces and cutting power consumption to some extent by softening or reducing the shear strength, τ_r of the work material ahead the cutting edge. To attain or enhance such benefit the work material ahead the cutting zone is often additionally heated externally.

This technique is known as Hot Machining and is beneficially applicable to the work materials which are very hard and hardenable like high manganese steel, Hadfield steel, Nihard, Nimonic etc.

2.4　Measurement of Tool Tip Temperature

To measure the temperature in the cutting zone, some methods have been developed. For that purpose, the work-tool thermocouple technique is the most widely used. The measurement system, described by Nakayama (1956), is shown in figure.

The test is first conducted without cutting and the reading on the Milli voltmeter resulting from the rubbing action of the constant wire on the work piece is recorded. This reading is subsequently subtracted from the readings taken while cutting was in progress.

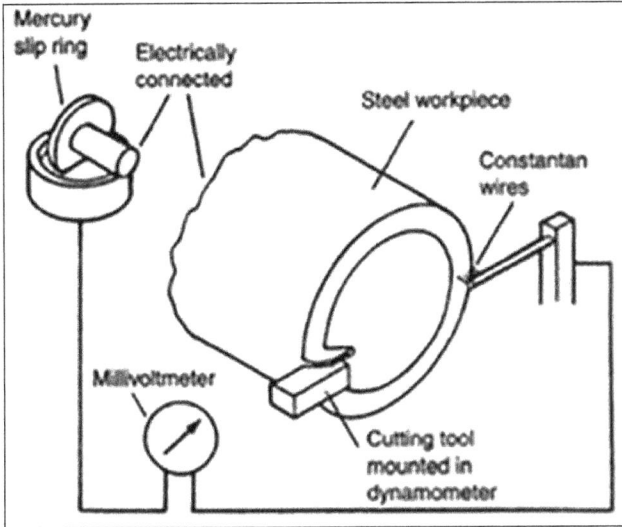

Using this method, the temperature at selected points around the end-face of the tubular work piece are measured and then used to calculate the proportion of the shear-zone heat conducted into the work piece.

When the tool work piece area can be observed directly, a camera with film sensitive to infrared radiation can be used to determine temperature distributions. In such a technique, a furnace of known temperature distribution was photographed simultaneously with the cutting operation using an infrared-sensitive plate, enabling the optical density of the plate to be calibrated against temperature.

Some other methods of temperature measurements include:

- Embedded-thermocouple technique.

- Metal microstructure and micro hardness variation measurements.

- Thermo sensitive painting technique.

- Temper-color technique.

Determination of Cutting Temperature

The magnitude of the cutting temperature need to be known or evaluated to facilitate:

- Assessment of machinability which is judged mainly by cutting forces and temperature and tool life.

- Design and selection of cutting tools.

- Evaluate the role of variation of the different machining parameters on cutting temperature.

- Proper selection and application of cutting fluid.

Analysis of Temperature Distribution in the Chip, Tool and Job

The temperatures which are of major interests are:

θ_s : Average shear zone temperature.

θ_f : Temperature at the work-tool interface (tool flanks).

θ_i : Average (and maximum) temperature at the chip-tool interface.

θ_{avg} : Average cutting temperature.

Cutting temperature can be determined by two ways:

- Analytically – Using mathematical models (equations) if available or can be developed. This method is simple, quick and inexpensive but less accurate and precise.

- Experimentally – This method is more accurate, precise and reliable.

- Analytical estimation of cutting temperature, θ_s.

(a) Average Shear Zone Temperature, θ_s

The cutting energy per unit time, i.e., P_z, V_C gets used to cause primary shear and to overcome friction at the rake face as,

$$P_Z.V_C = P_S.V_S + F.V_f$$

where,

V_S = Slip velocity along the shear plane,

and V_f = Average chip – Velocity,

So, $P_S.V_S = P_Z.V_C - F.V_f$.

Equating amount of heat received by the chip in one minute from the shear zone and the heat contained by that chip, it appears,

$$\frac{A.q_1\left(P_z \cdot V_c - F \cdot V_f\right)}{J} = C_V a_1 b_1 c_1 V_c \left(\theta_s - \theta_a\right)$$

where,

A = Fraction (of shear energy that is converted into heat).

q_1 = Fraction (of heat that goes to the chip from the shear zone).

C_v = Volume specific heat of the chip.

J = Mechanical equivalent of heat of the chip / work material.

θ_a = Ambient temperature.

$a_1.b_1$ = Cross sectional area of uncut chip,

$= ts_o$

Therefore, $\theta_s = \dfrac{Aq_1\left(P_z \cdot V_c - F \cdot V_f\right)}{Jts_o\, V_c} + \theta_a$

Or $\theta_s \cong \dfrac{Aq_1\left(P_z - F/\zeta\right)}{Jts_o}$

Generally A varies from 0.95 to 1.0 and q from 0.7 to 0.9 in machining like turning.

(b) Average Chip – Tool Interface Temperature, θ_i

Using the two dimensionless parameters, Q_1 and Q_2 and their simple relation (Buckingham):

$$Q_1 = C_1 \cdot Q_2^n$$

Where, $Q_1 = \left(\dfrac{C_v\theta_i}{E_C}\right)$ and $Q_2 = \left(\dfrac{V_c c_v a_1}{\lambda}\right)^{0.5}$

E_C = Specific cutting energy

λ = Thermal conductivity

c_v = Volume specific heat

c_1 = Constant

n = Index close to 0.25

Therefore, $\theta_i = c_1 E_C \sqrt{V_c a_1 / \lambda C_v}$ ...(i)

Using equation (i), one can estimate the approximate value of average θ_i from the known other machining parameters.

Turning, Shaping and Planing Machines

3.1   Turret Lathe and Capstan Lathe

Main Parts of a Turret Lathe

1. Turret head and saddle: A turret is a block which can hold number of tools at time. The turret can be indexed about a vertical axis. Generally, a hexagonal turret is used. It has six faces. On each face, there is a bore to receive the shank of a tool. Four tapped holes are available on each face of the turret for clamping tool holders.

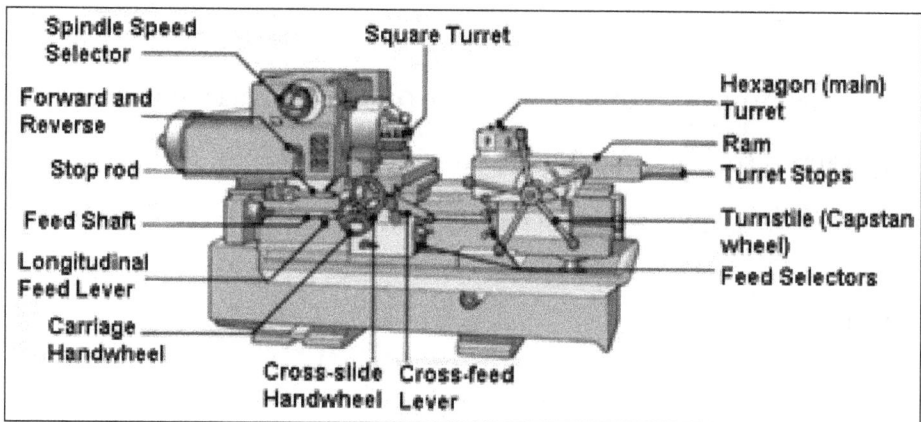

Capstan and turret lathes.

2. Cross Slide: The cross slide is provided over the bed of the lathe between the head stock and turret saddle. The cross slide has two tool posts. One tool post is mounted at the front of the machine. This front tool post has a square turret. This has four faces for mounting the cutting tools.

Working Principle of Turret Lathe

It consists of following main parts:

- Head stock

- Carriage or Saddle

- Turret saddle

- Bed
- Legs

Head Stock

A capstan and turret lathe carries a similar type of head stock as in a center lathe, but it is comparatively larger in size and heavier in construction in order to provide a wider range of speeds.

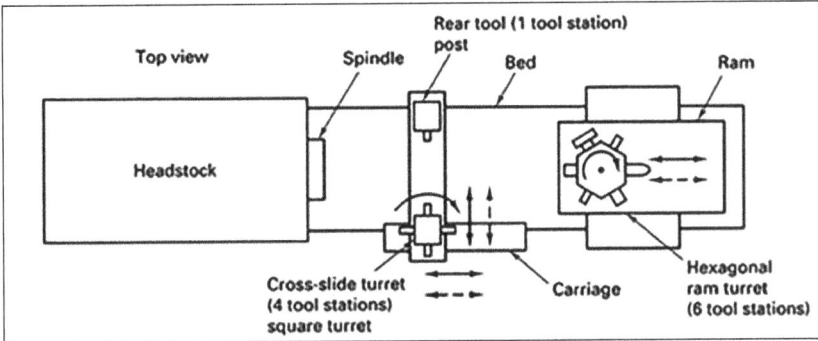

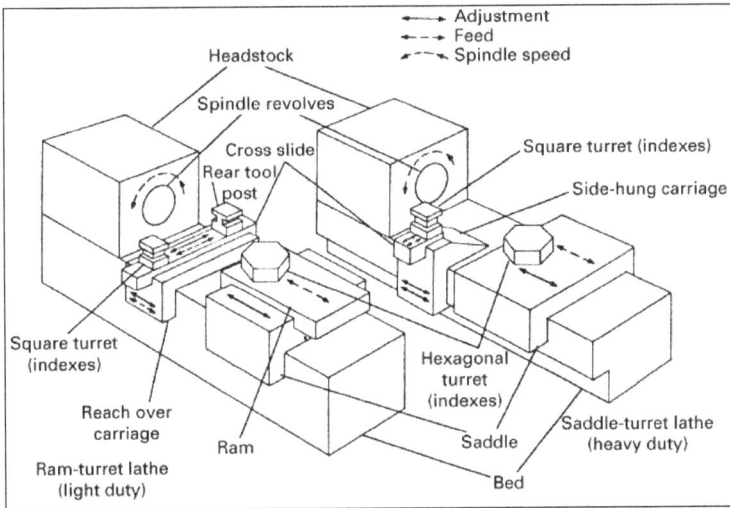

Block diagram of a turret lathe (Top view).

The following types of head stocks are commonly used in the process:

- Cone pulley type.
- Direct motor driven.
- All geared head stock.

Carriage

The carriage or saddle mounted on the bed of a turret lathe is almost similar to the saddle of an engine lathe. It carries a cross-slide over it, on which two tool-posts are mounted. One at the front and other at the rear. By using a handle provided at the top of the tool post, the tool can be indexed in the required position.

Turret Saddle

Turret saddle replaces the tail stock of a center lathe. It is mounted directly on the lathe bed on the same side as a tail stock in case of a center lathe. It is provided with a slide which moves in the guide ways made in it and the turret head is mounted on the slide. During the operation, it remains stationary and the tools are fed longitudinally by moving the slide.

The turret head mounted on the saddle is hexagonal in turret lathe which carries six holes, one on each flat face or equally spaced along the periphery of the head.

Bed

Bed is a box type casting. It consists of stiffener, longitudinal ribs and provided with parallel guide ways over its top to enable sliding of carriage and, turret saddle over them. Due to parallel guide ways, all the sliding parts slide in perfect alignment.

Legs

Each lathe carries two legs, one below each end of the bed. These legs are hollow casting which bear the entire load of the bed, sliding and stationary parts mounted on the bed, tooling and work-holding devices.

Tool Layout of a Turret Lathe

A tool layout is prepared for the manufacture of square headed bolt from a square bar stock.

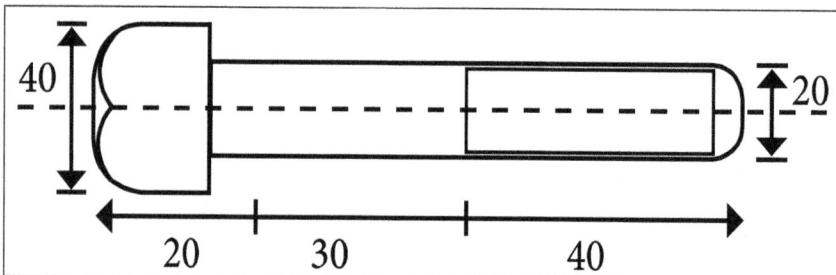

Using a turret lathe as shown in figure.

Stage-1:

- The component drawing is drawn.

- The tool length of the work is calculated and 10 mm is added to provide clearance.

- The number of operations involved is roughly listed.

- The sequence of operation is assigned.

- The proper machine of 75mm turret lathe is selected.

- The proper material of mild steel square bar is selected.

- All the tools and equipment's as per operation sequence are collected and fitted on turret faces or on cross-slides as per our convenience.

Stage-2:

The tool layout is drawn as shown in figure.

[Note: Number tools fitted in the turret face are only four. So for providing uniform balancing, tools are arranged like the first two are in successive faces and other two are in next successive faces by leaving one lace left to free].

Stage 3:

- Tooling schedule chart (to machine square bolt).

- Machine: 75mm turret lathe.

- Material: Square mild steel bar.

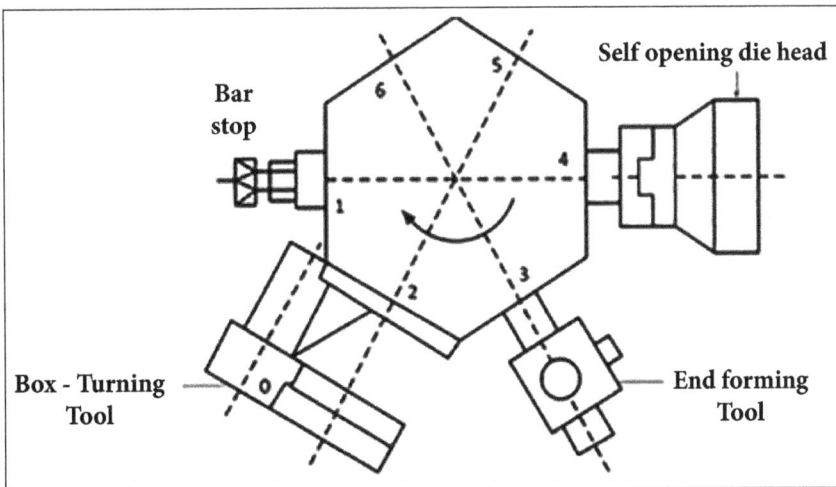

Sequence of operation tool.

Description of operation sequence	Tool position	Tool
1. Holding the square bar in collet and setting the required length of 100mm (90 + 10).	Turret positions.	Bat stop.

2. Turn to 20 mm. diameters to a length 70mm (from the right end).	Turret positions-2.	Roller steady box.
3. Form the right end of the bolt.	Turret positions-3.	Roller steady bar Ending tool.
4. Make the external thread cutting of 20mm diameter to a length of 40mm (from the right end).	Turret positions-4.	Self-opening die head with chases of 20mm.
5. Chamfering the bolt head.	Front cross slid tool position-1.	Chamfering tool.
6. Parting off the work piece.	Rear tool positions.	Parting tool is place in inverted position (making the rotation of work anticlockwise with respective to tool).

- Setting the bar stop: The bar stop (1) is set at the distance of 100mm from the collet face by using slip gauge. An extra length of 10mm is allowed for parting off (4mm) and clearance of the collet face (6mm). This clearance is allowed to penetrate the parting tool deep into the work piece without any Inference.

- Setting of the roller steady box turning rod: This tool is set on turret face of 2. This tool (2) is used for turning the work piece to 200mm diameter and 80cm long from the right end.

- Setting of bar ending tool: This tool is set (3) on fourth turret face but turret position-3. This is used to chamfer the right end of the work piece.

- Setting self-opening die head: This tool (4) is set on the fifth face of the turret. The proper blades of chasers are selected and fitted into the die head to cut a thread of 20mm diameter.

- Setting of Chamfering tool: This tool (5) is set on the cross-slide front-end position-1 used to chamfer the bolt head edges by giving cross feed.

- Setting of parting tool: This tool is set on the rear end of cross-slide. It is used to part the work piece after completing all operations.

Advantages of Turret Lathe

- Larger and heavier chucking works are usually handled on a turret lathe, whereas a capstan lathe is suitable for bar work.

- In the case of turret lathe, the turret is mounted on the saddle which slides directly on the lathe bed ways. This type of construction provides utmost rigidity to the tool support of the entire cutting load, which is taken up by the lathe bed directly.

- Heavier turret lathes are equipped with power chucks like air operated chucks for holding larger sizes of work quickly.

Turret Indexing Mechanism

A simple line sketch of the mechanism is shown in figure, it illustrates an inverted plan of the turret assembly. The turret (1) is mounted on the spindle (5), which rest on the bearing on the turret saddle. The index plate (2) the bevel gear (3) and an indexing ratchet (4) are keyed to the spindle (5).

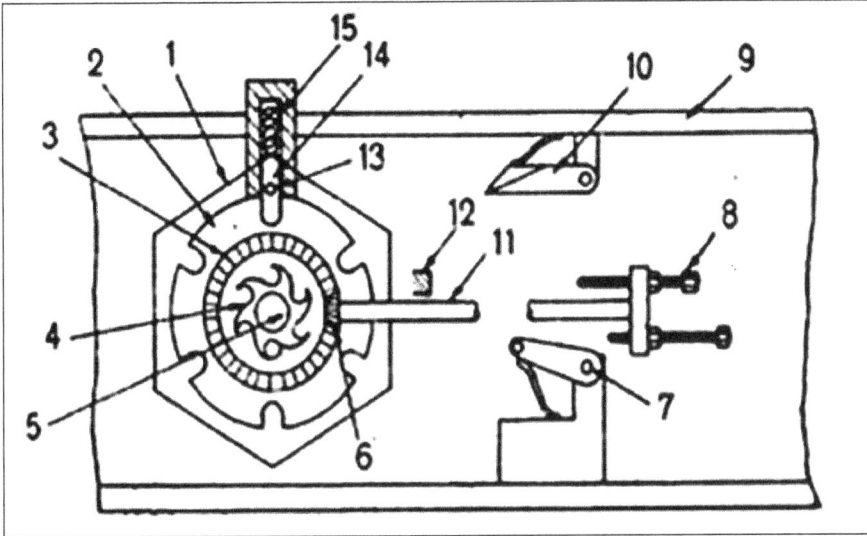

Turret indexing mechanism.

1. Hexagonal turret

2. Index plate

3. Beveled gear

4. Indexing ratchet

5. Turret spindle

6. Beveled pinion

7. Indexing pawl

8. Screw stop rods

9. Lathe bed

10. Plunger actuating cam

11. Pinion shaft

12. Stop

13. Plunger pin

14. Plunger

15. Plunger spring

The plunger (14) fitted within the housing and mounted on the saddle locks the index plate by spring (15) pressure and prevents any rotary movement of the turret as the tool feeds into the work. A pin (13) fitted on the plunger (15) projects out of the housing. An actuating cam (10) and the indexing pawl (7) are spring loaded.

As the turret reaches the backward position, the actuating cam (10) lifts the plunges 14 out of the groove in the indexing plate due to the riding of the pin (13) on the beveled surface of the cam (10) and thus unlocks the index plate (2). The spring loaded pawl (7) which by this time engages with a groove of the ratchet plate (4) causes the ratchet to rotate as the turret head moves backward.

Bar Feed Mechanism

The capstan and turret lathe while working on bar work require some mechanism for bar feeding.

The long bar which protrude out of the head stock spindle requires to be fed through the spindle up to the bar stop after the first piece is completed and the collet chuck is opened.

In simple cases the bar may be pushed by hand, but this process unnecessarily increases the total time, because the spindle and the large bar must come to a deal stop before any adjustment can be made. Thus in each case unnecessarily long time is wasted in stopping, setting and starting the machine.

Various types of bar feeding mechanisms have therefore been designed, which push the bar forward immediately after the collet release the work without stopping the machine, enabling the setting time to be reduced to the minimum. Figure shows the simple bar feeding mechanism.

The bar (6) is passed through the bar chuck (3), spindle of the machine and then through the collet chuck. The bar chuck (3) rotates in the sliding bracket body (2) which is mounted on long slide bar.

The bar chuck (3) grips the bar centrally by two set screws (5) and rotates with the bar in the sliding bracket body (2). One end of the chain (8) is connected to the pin (9) fitted on the sliding bracket (10) and the other end supports a weight (4), the chain running over two pulleys (7) and (11) mounted on the slide bar.

The weight (4) constantly exerts end thrust on the bar chuck while it revolves on the sliding bracket and forces through the spindle, the moment the collet chuck is released. Thus the bar feeding may be accomplished without stopping the machine.

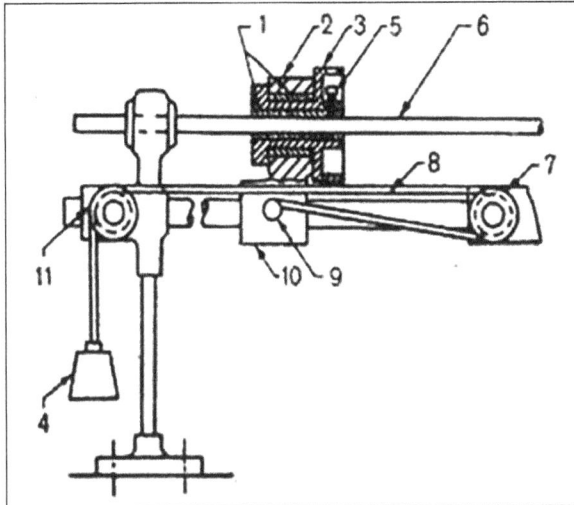

Bar feeding mechanism.

1. Chuck bush

2. Sliding bracket body

3. Bar chuck

4. Weight

5. Bar chuck set screw

6. Bar

7. Pulley

8. Chain

9. Pin on the sliding bracket

10. Sliding bracket

11. Pulley

Line-diagram of a Capstan Lathe and its Principal Parts

The mains parts of capstan lathe are:

- Bed

- Cross slide

- Head stock

- Turret head.

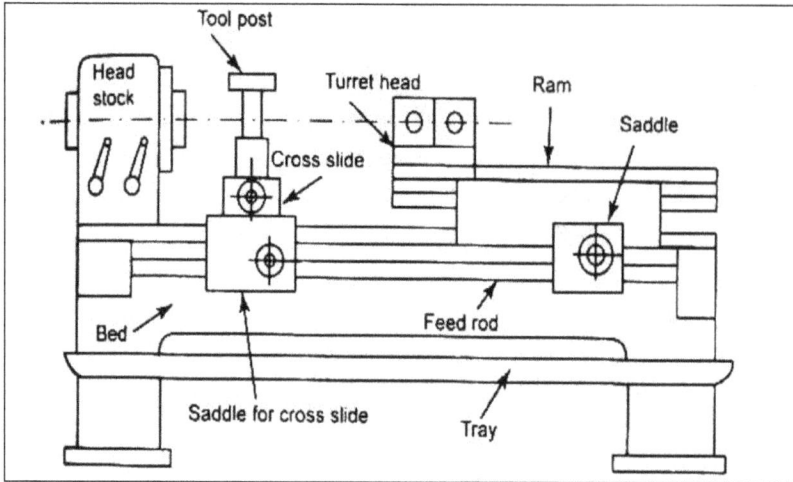

Capstan and Turret Lathe.

Bed

Bed is the base part of the lathe. It is a box type which is made of cast iron. Guide ways on the top of the bed has been provided accurately. Cross slide and turret head are mounted on these guide ways. The bed should be strong and withstand heavy loads, force and vibrations during machining task.

Cross Slide

The two types of cross slides are:

- Reach over type
- Side hung type

(i) Reach over type: It is mounted on the bed guide ways in between head stock and turret which is also supported by lower rail. The cross slide has two tool posts. One is at the front end having four faces of square turret for mounting the tools. Each tool can be indexed by 90°.

(ii) Side hung type: This type of cross slide is entirely supported on the front way which has no rear tool post. This provides greater swing capacity to accommodate large diameter work piece. It is mainly used in turret lathe.

Head Stock

Head stock of capstan and turret lathe is similar to that of head of ordinary center lathe but larger and heavier in construction to house the spindle and driving mechanism. A powerful motor of 30 to 2000rpm speeds is fitted.

Turret Head

A turret head has a hexagonal block having six faces with a bore for mounting six or more than six tools at a time. The four threaded holes on these faces are used to accommodate the tool holders. The turret head is mounted on the ram fitted with turret slides longitudinally on a saddle.

Legs

Each lathe carries two legs. These legs are hollow casting which bear the entire load of the bed, sliding and stationary parts mounted on the bed, tooling and work-holding devices, etc.

Holding Devices in A Lathe

In capstan and turret lathes, work is not held between centers and there is no tail stock to support the work. The work is held in chucks or fixtures only.

Collect chuck: Small components are produced in large number in capstan lathe generally these components are produced from work pieces in the form of bar stock.

The different types of collets used in capstan lathes are:

- Draw back collet

- Push out collet

- Dead length collet

(i) Draw back collet: This is also known as draw in type collet here the taper of the collet nose and sleeves converge towards the left. To grip the work, the collet is pulled back by the collet tube into the tapper bore of the hood or sleeve.

(ii) Push out collet: Here the taper of the collet nose and sleeve converge towards the right. In this type, the tapered portion of the collet is pushed to the right.

(iii) Dead length collet: In this type, there is no axial movement of the collet. The taper of the collet nose and sleeves converge towards the left, a sliding sleeve is placed between the collet and the hood.

Jaw Chucks

Jaw chucks that are used in a center lathe are also used in turret lathes .Jaw chucks are used for holding jobs of larger size.

- Three jaw self-centering chuck.

- Four jaw independent chuck.

- Air operated chuck.

(i) Three jaw chuck: These type chucks are used for holding cylindrical or hexagonal shaped works, the jaws are moved simultaneously inward or outward by means of a single key.

(ii) Four jaw chuck: These chucks are used for holding rough and irregular shaped components. It can also hold square or octagonal components easily.

(iii) Air operated chuck: These chucks are used for heavy duty works. The chuck grip the work quickly and with more force the jaws are actuated by air pressure. An air cylinder is fitted at the back of the head stock.

Comparison between Capstan Lathe from Turret Lathe

S. No.	Turret Lathe	Capstan Lathe
1.	Its turret (bead) U is mounted directly on the saddle.	Its turret (bead) is mounted on an auxiliary Slide, which moves on the guide ways provided on the saddle.
2.	For feeding the tools to the work the entire saddle unit is moved.	In this case the saddle is fixed at a convenient distance from the work and the tools are fed by moving the slide.

3.2 Tool Layout

The cutting time for a given operation is mainly controlled by proper tooling, speed and feed. Much time is saved by taking combined or multiple cuts. In bar work, combined cuts provide additional support to the work and eliminate springing action and chatter.

The method of tooling and the sequence of operations for making internal threads on a component are shown in the figure (1) below.

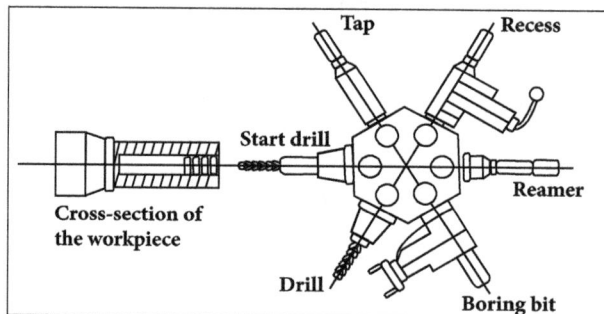

(1) Tooling layout of the threaded component having internal thread.

It involves the following steps:

- Advance the bar stock against the combined stock stop. Locate the proper position with the drill and clamp the job in the collet. Advance the start drill in the job after centering the work piece.

- Drill the job to the required length.

- Bore the thread diameter to the required size.

- Ream the diameter to exact sizes.

- Recess a groove for thread clearance. The operation is performed by a quick acting slide tool mounted on the boring bar.

- Cut the threads with the tap.

Part Off the Job with a Parting Tool

Layout for a ball bearing part: The sectional view of a ball bearing part is shown in the figure (2). We are interested in the production of such parts on a turret lathe. The typical tool layout is shown in the figure (3).

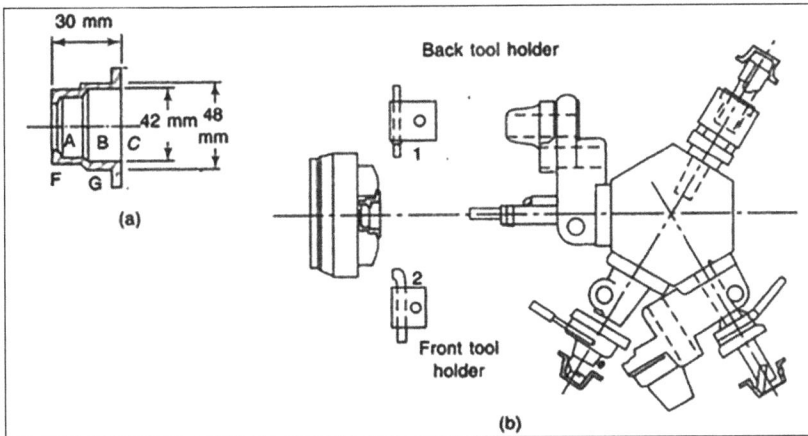

(2) (a) Ball bearing part (b) Layout of ball bearing part.

It involves the following steps:

- Rough face the bar end with the tool held in the back tool post.

- Finish the face end of the part with the tool held in the front tool post.

- Bore diameters A and B and chamfer C with tools fitted in the boring bars and held in the hexagonal turret.

- Recess diameters F and G with the help of a double recessing cutter held in the re-cessing tool slide.

- Bore diameter A to size using a fine adjustment boring tool.

- Generate threads on bore B using tap set.

- Part off the work piece with a parting tool.

Layout for front wheel axle The sketch of the front wheel axle to be produced on a capstan lathe is shown in the figure (3a).

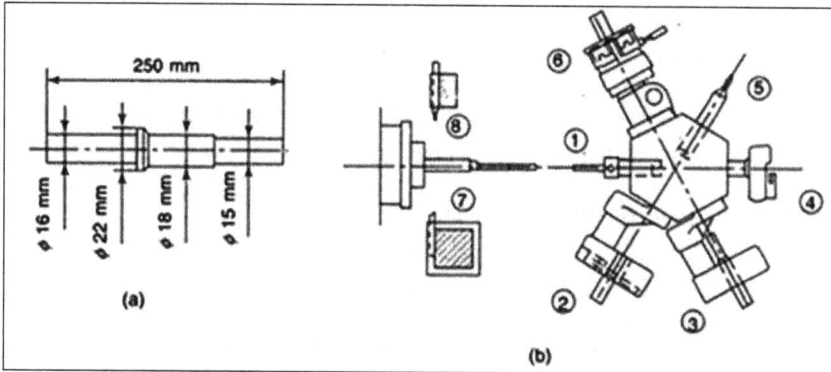

(3) (a) Front wheel axial (b) Layout of front wheel axile.

The material required for its manufacture is a 22 mm steel bar. The process of production of components involve the following steps:

- Lay all the tools as shown in the figure (3b).

- Feed out the adjustable.

- Turn 15 mm diameter a with a box tool.

- Turn 18 mm diameter with a box tool.

- Chamfer the end with a chamfering tool.

- Mark the centre with a centre drill.

- Cut the external threads with a die.

- Form a 22 mm diameter.

- Chamfer with the tool in the turret.

- Make a 16 mm diameter.

- Part off the component.

3.2.1 Shaping Machine

- Base: The base is of cast iron is grounded to the workshop floor.

- Column: Column is a box type casting mounted on the base, it houses the ram driving mechanism. Guide ways are provided on the top of column, for the ram to reciprocate.

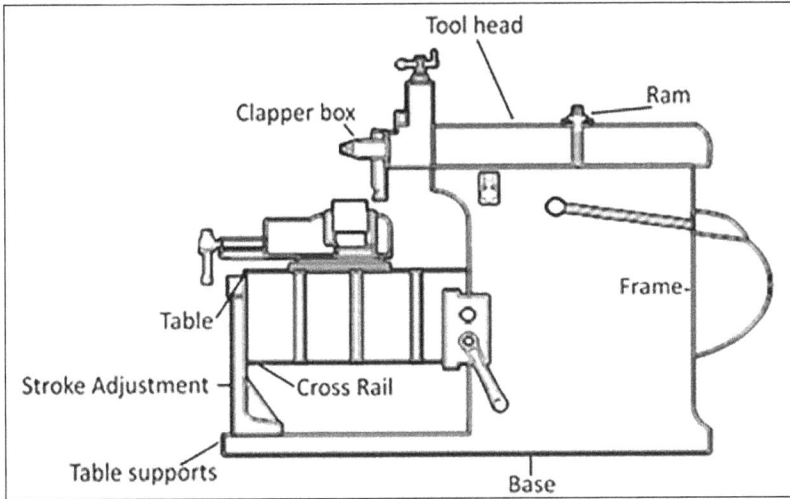

Shaper machine.

- Cross rail: Cross rail is guided by the vertical guide ways of the column in front of the cross rail, two horizontal guides are provided for the saddle to slide along.

- Table: Table is a box type casting. The front face of the table is supported by an adjustable support and the rear face is attached to the saddle. Its top and sides are perfectly machined and the sides are square with the top surface.

- Ram: This is the reciprocating member which carries the tool head in the front end. Ram is semi -cylindrical in shape.

- Tool head: Tool head provides vertical and angular feed movement of the tool. Tool head consists of a vertical tool slide, an apron and a tool holder.

By rotating the tool head on the graduated swivel base, tool slide can be turned to the required angle. This is required to give angular feed to the machine surfaces at an angle.

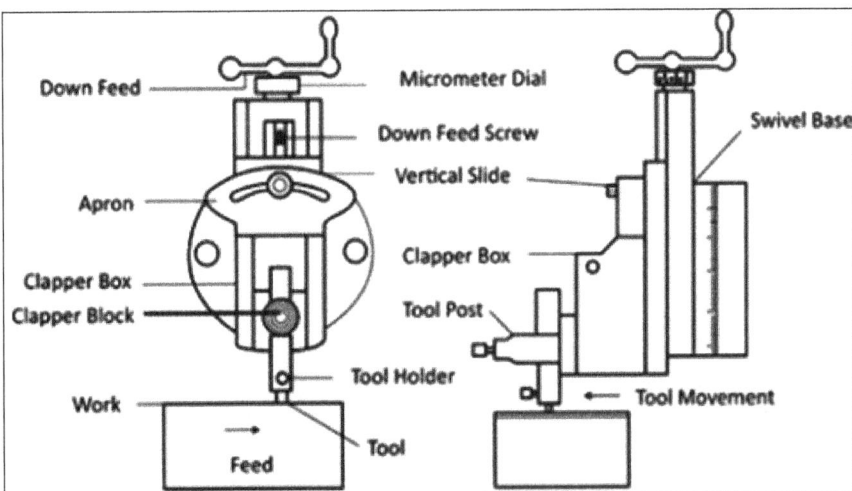

Tool head.

Feed Mechanism

In a shaper both down feed and cross feed movements are provided intermittently and during the end of return stroke only. Vertical or bevel surfaces are produced by rotating the down feed screw of the tool head by hand. Cross feed movement is used to machine a flat horizontal surface. This is done by rotating the cross feed screw either by hand or power.

Rotation of the cross feed screw causes the table mounted upon the saddle to move sideways through a predetermined amount.

Figure below shows, the automatic cross feed mechanism of a shaper. The rotation of the bull gear causes the driving disc (8) to rotate in a particular direction. The driving disc (8) is T slotted and position of the crank pin (9) attached to the connecting rod may be altered to give different throw of eccentricity.

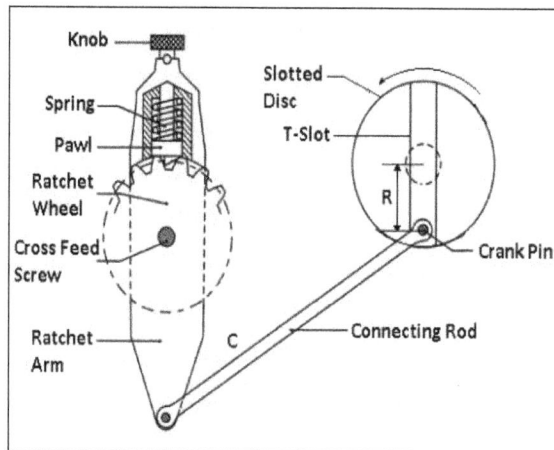

Automatic feed mechanism of a shaper.

1. Knob

2. Pin

3. Helical spring

4. Pawl

5. Ratchet wheel

6. Rocker arm fulcrum

7. Rocker arm connecting pin

8. Driving disc

9. Crank pin

The other end of the connecting rod is attached to the rocking arm by a connecting pin (7). The rocking arm is fulcrum at (6), the center of the ratchet wheel (5). The ratchet wheel (5) is keyed to the cross feed screw. The rocking arm houses a spring loaded pawl (4), which is straight on one side and bevel on the other side.

As the driving disc rotates, the connecting rod starts reciprocating and the rocking arm rocks on the fulcrum (6). When the driving disc rotates through half of the revolution in the clockwise direction, top part of the rocking arm moves in the clockwise direction and the pawl (4) being slant on one side slips over the teeth of the ratchet wheel (5) imparting it no movement.

As the driving disc rotates through the other half, the top of the rocking arm now moves in the anticlockwise direction and the straight side of the pawl engages with the teeth of the ratchet wheel causing the wheel to move in anticlockwise direction only.

As the driving disc is connected to the bull gear the table feed movement is effected when the bull gear or the driving disc rotates through half of the revolution, i.e., during return stroke only. Rotation through other half imparts no feed movement.

To reverse the direction of rotation of ratchet wheel and consequently the feed, a knob (1) on the top of the pawl (4) after removing the pin (2) is rotated through 180°. The amount of feed may be altered by shifting the position of crank pin 9 with respect to the centre.

Work Holding Devices

The work may be supported on the table by the following methods depending on the nature of the work piece.

- Clamped in a vise.
- Clamped on the table.
- Clamped to the angle plate.
- Clamped on a V-block.
- Held between shaper index centre.

Shaper Tools

The cutting tool used in a shaper is a single point cutting tool having rake, clearance and other tool angles similar to a lathe tool.

It differs from a lathe tool in tool angles.

Shaper tools are much more rigid and heavier to withstand shock experienced by the cutting tool at the commencement of each cutting stroke.

In a lathe tool the effective angle of rake and clearance may be varied by raising or lowering the point of the tool in relation to the centre of the work, but in a shaper the tool angles cannot be changed as the tool is always clamped perpendicular to the surface of the work.

As the tool removes metal mostly from its side cutting edge, side rake of 10° is usually provided with little or no front rake.

The side rake angle to be provided is dependent upon the kind of metal being cut, the hardness of the tool material, type of cut and other factors which influence the rake angle.

A shaper can also use a right hand or left hand tool. The left hand tool is more common because it gives the cut better than the right hand tool.

High speed steel is the most common material for a shaper tool but shock resistant cemented carbide tipped tool is also used where harder material is to be machined.

Specifications of Shaping Machines

- Maximum length of stroke or cut it can make.
- Maximum horizontal travel of table.
- Maximum vertical travel of table.
- Size of the table.
- Type of drive system.
- Floor space required.
- Cutting to return stroke ratio.

3.3   Planing Machine

Planer is a very large reciprocating machine tool. The work is mounted on the table by any one of the work holding devices. Two vertical columns with vertical guide ways are provided on both sides of the bed and connected by a cross-rail to mount the tool heads and also connected by a cross beam at the top. These tool heads are used to hold the tools. The tool cuts the work piece when the table reciprocates.

The double planer has the following parts:

- Bed
- Table
- Columns
- Cross rail

- Tool head

Bed

The bed is a very strong and rigid of box type which is made by casting process. The best length is made twice the length of table with 'V' guide ways. The table is mounted over the bed which houses various mechanisms.

Table

It is also a box type structure which reciprocates on the bed guide ways. It is also having 'T' slots.

Columns

The two long structured member along with guide ways are provided on both sides of the member. The two long columns are connected by a cross rail and cross beam. The cross rail slides on these guide ways. It carries feed mechanism and power transmission links.

Cross Rail

It is a rigid structural member mounted between two columns and slides on the guide ways already provided on the columns. The cross rails can be set or clamped at any height. It carries tool heads.

Tool Heads

Maximum four tool heads can be mounted on the planer. Two are on the cross rail and another two are on the guide ways of both the columns. It may tilt to any required angle.

Types of Planer

Like shapers, planers are also basically used for producing flat surfaces. But planers are very large and massive compared to the shapers. Planers are generally used for machining large work pieces which cannot be held in a shaper. The planers are capable of taking heavier cuts. The planer was first developed in the year 1817 by Richard Roberts, an Englishman.

The different types of planer which are most commonly used are:

- Standard or double housing planer.
- Open side planer.
- Pit planer.

- Edge or plate planer.

- Divided or latching table planer.

Double Housing Planer

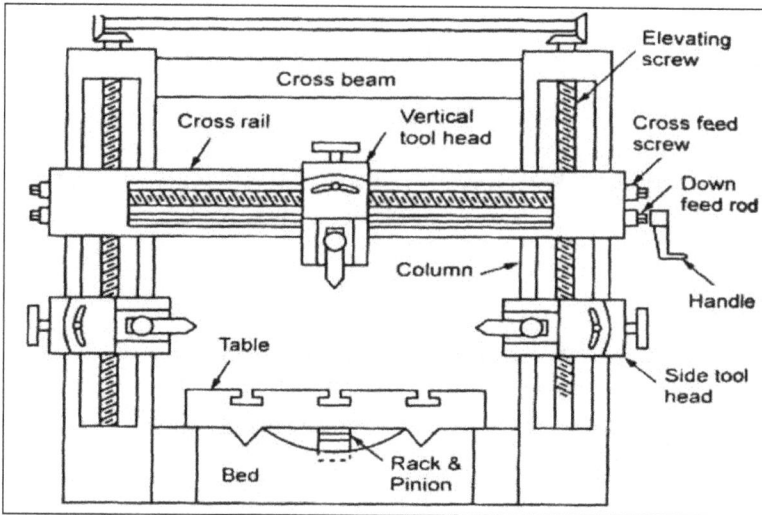

Schematic view of a double housing planer.

It is most widely used in workshops. It has a long heavy base on which a table reciprocates on accurate guide ways. It has one drawback. Because of the two housings, one on each side of the bed, it limits the width of the work that can be machined.

Open Side Planer

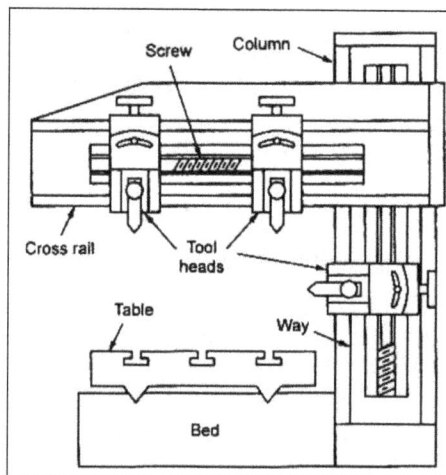

Schematic view of an Open side planer.

It has a housing only on one side of the base and the cross rail is suspended from the housing as a cantilever. This feature of the machine allows large and wide jobs to be clamped on the table. As the single housing has to take up the entire load, it is made extra-massive to resist the forces. Only three tool heads are mounted on this machine. The constructional and driving features of the machine are same as that of a double housing planer.

Pit Planer

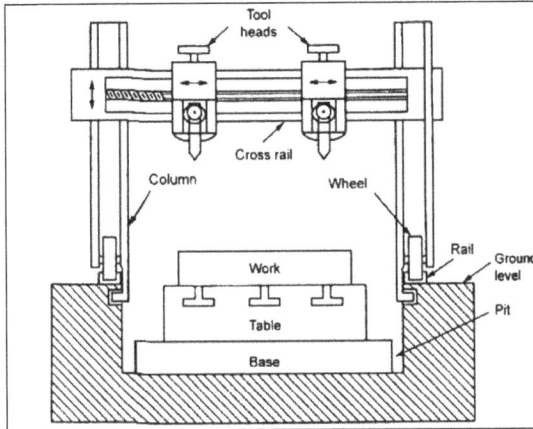

Schematic view of Pit planer.

It is massive in construction. It differs from an ordinary planer in that the table is stationary and the column carrying the cross rail reciprocates on massive horizontal rails mounted on both sides of the table.

This type of planer is suitable for machining a very large work which cannot be accommodated on a standard planer and the design saves much of floor space. The length of the bed required in a pit type planer is little over the length of the table.

Edge or Plate Planer

The design of a plate or edge planer is totally unlike that of an ordinary planer. It is specially intended for squaring and beveling the edges of steel plates used for different pressure vessels and shipbuilding works.

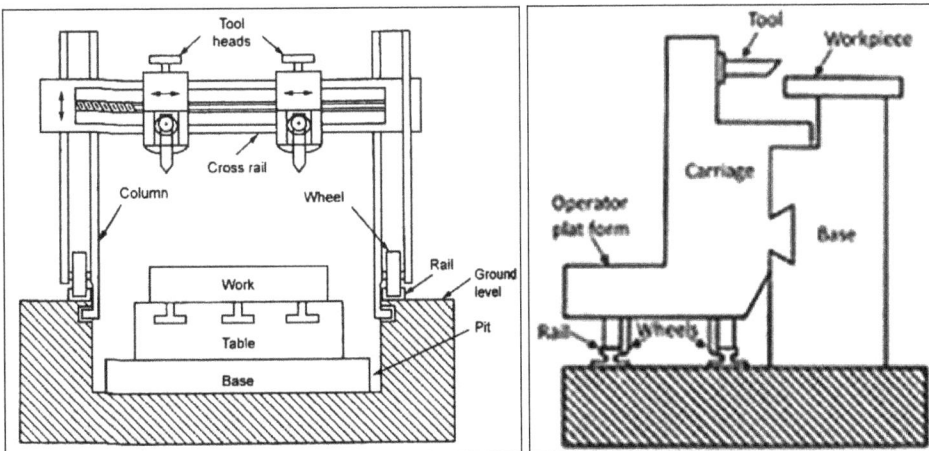

Schematic view of Edge or plate planer.

Divided Table Planer

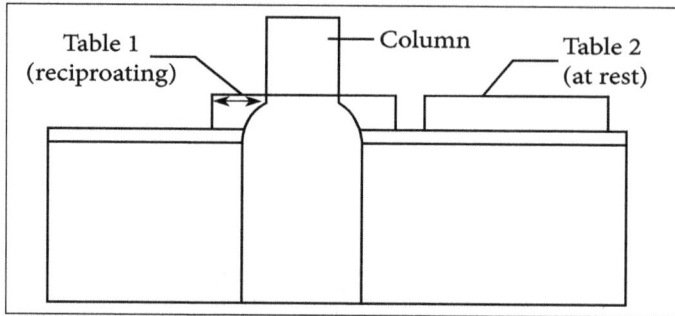

Schematic view of Divided table planer.

This type of planer has two tables on the bed which may be reciprocated separately or together. This type of design saves much of idle time while setting the work. To have a continuous production one of the tables is used for setting up the work and the other is used for machining. This planer is mainly used for machining identical work pieces. The two sections of the table may be coupled together for machining long work.

Comparison of Shaper and Planer

S. No.	Shaper	Planer
1.	Tool reciprocates and the work is stationary.	Tool is stationary and work reciprocates.
2.	This machine is used for machining medium and small work pieces.	It is used for machining large and heavy work pieces.
3.	Less accuracy due to the overhanging of ram supported during cutting.	It gives more accuracy as the tool is rigidity.
4.	Production time is more since it has one tool head.	Production time is less since it has two or four tool heads.
5.	Work setting is easier.	Work setting requires more skill.
6.	Heavy cut cannot be given.	Heavy cut can be given as it has rigid base and uses strong tools.

3.4   Driving Mechanisms of Lathe

Hydraulic mechanism is used as a quick return mechanism.

- The shaper ram in the position as shown moves from left to right.

- The oil from the reservoir is drawn through the filter by a gear pump which is driven by an electric motor.

- Pump delivers a constant quantity of oil to the control valve and from control valve, the oil can be delivered to the either side of the piston in the cylinder situated under the ram.

- To which side of the piston the oil will be delivered is determined by the position of the control lever of the control valve.

- Figure shows a constant volume hydraulic mechanism.

- The same volume of oil is delivered on both sides of the piston, but the intensity of pressure is different due to difference in effective area of the piston.

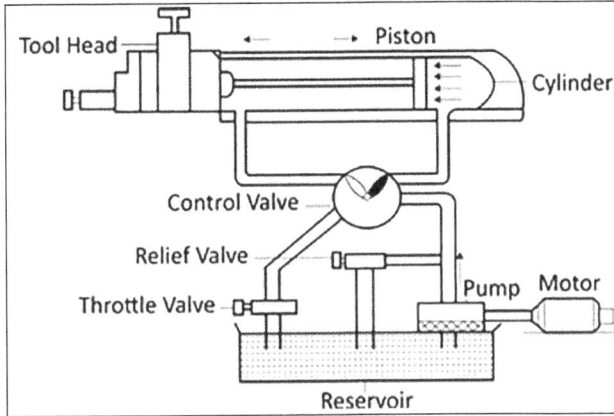

Hydraulic mechanism.

- Due to difference in pressure, the ram travels at a faster rate in the return stroke than in the forward stroke.

- The stroke length of the ram is adjusted by adjusting the distance between the two stops.

- The cutting speed is controlled by using the throttle valve during the cutting stroke.

3.4.1 Shaper Machine and Planing Machine

In a shaper, rotary movement of the drive is converted into reciprocating by the mechanism within the column of the machine. The ram holding the tool gets the reciprocating movement. In a standard shaper metal is removed in the forward cutting stroke, while the return stroke goes idle and no metal is removed during this period. To reduce the total machining time it is necessary to reduce the time taken by the return stroke. The shaper mechanism should be designed that it can allow the ram holding the tool to move at a comparatively slower speed during the forward cutting stroke.

During the return stroke it can allow the ram to move at a faster rate to reduce the idle return time. This mechanism is known as quick return mechanism. The reciprocating movements of the ram and the quick return mechanism of the machine are usually obtained by anyone of the following methods:

- Crank and slotted link mechanism.

- Whitworth quick return mechanism.

- Hydraulic shaper mechanism.

Crank and Slotted Link Mechanism

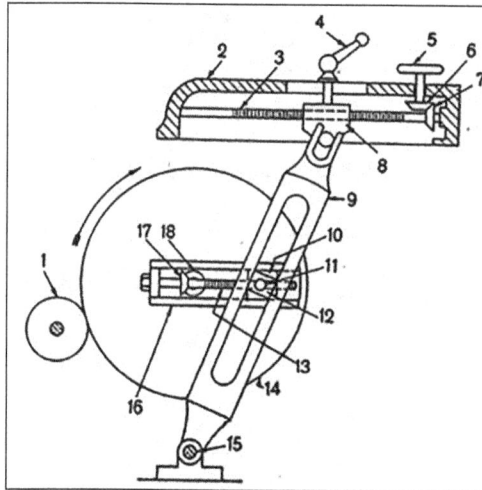

Crank and slotted link mechanism.

1. Driving pinion

2. Ram

3. Screwed shaft

4. Clamping lever

5. Hand wheel for position of stroke adjustment

6. Bevel gears

7. Ram block

8. Slotted link or rocker arm

9. Bull gear sliding block

10. Crank pin

11. Rocker arm sliding block

12. Lead screw

13. Bull gear

14. Rocker arm pivot

15. Bull gear slide

16. Bevel gears

The motion or power is transmitted to the bull gear (14) through a pinion (1) which receives its motion from an individual motor. Speed of the bull gear may be changed by different combination of gearing or by simply shifting the belt on the step cone pulley.

Bull gear is a large gear mounted within the column. Bolted to the centre of the bull gear is a radial slide (16) which carries a sliding block (10) into which the crank pin (11) is fitted. Rotation of the bull gear will cause the crank pin to receive a uniform speed. Sliding block (12) which is mounted upon the crank pin is fitted within the slotted link (9). The slotted link also known as rocker arm is pivoted at (15) at its bottom end attached to the frame of the column.

The upper end of the rocker arm is forked and connected to the ram block (8) by a pin. As the bull gear rotates causing the crank pin to rotate, the sliding block (12) fastened to the crank pin will rotate on the crank pin circle and at the same time will move up and down the slot in the slotted link (9) giving it a rocking movement which is communicated to the ram. Thus the rotary motion of the bull gear is converted to reciprocating movement of the ram.

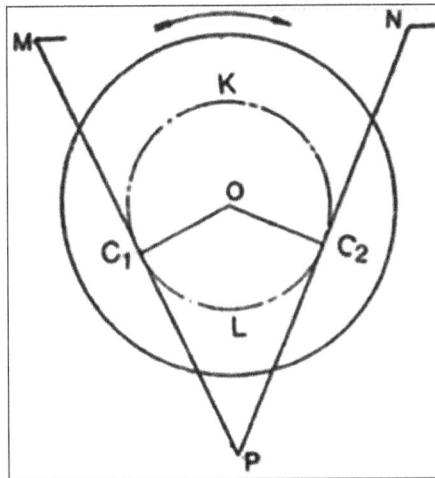

Principle of quick return mechanism.

When the link is in the position PM, the ram will be at the extreme backward position of its stroke and when it is at PN, the extreme forward position of the ram will have been reached. PM and PN are drawn target to the crank pin circle.

The forward cutting stroke takes place, when the crank rotates through the angle C_1KC_2 and the return stroke takes place when the crank rotates through the angle C_2LC_1. It is evident that the angle C_1KC_2 made by the forward or cutting stroke is greater than the angle C_2LC_1 described by the return stroke. The angular velocity of the crank pin being constant the return stroke as therefore, completed within a shorter time for which it is known as quick return motion.

Planer Driving Mechanism

A planer driving mechanism provides the longitudinal to and fro motion of the planer worktable. The following methods are employed for the driving mechanism:

- Open and cross belt drive.
- Reversible motor drive.
- Hydraulic drive.

Open and Cross Belt Drive

A driving mechanism of a planer consists of an electric motor situated over the housing. The motor shaft is coupled with a counter shaft. The counter shaft, at its extreme end, carries two driving pulleys: One for open belt and other for cross belt.

The main driving shaft is provided below the bed. Its one end passes through the housing and carries a pinion, which meshes with the rack provided under the table of the machine. The other end of the shaft carries two pairs of pulleys, each pair consists of a fast pulley and a loose pulley. One of these pairs is connected to one of the driving pulleys by means of an open belt and the other to the second driving pulley by means of crossed belt.

A speed reduction gear box is mounted on the main driving shaft and same is incorporated between the pinion and the pairs of driven pulleys. The driving mechanism is shown in the below figure. One set of the above pulleys is used for forward motion and other set for backward motion of the table.

The cross belt is used for forward motion and open belt is used for backward motion. The driving pulley on counter shaft for cross belt is smaller than the pair of fast and loose pulleys for the same. While the driving pulley on the driving shaft for open belt is bigger than the pair of fast and loose pulley on the same.

This arrangement is provided for slow forward stroke and Fast backward stroke. The pulleys are so arranged that when the cross belt is on fast pulley, i.e., in forward stroke, the open belt will be on the loose pulley and its reverse will take place during return stroke. The relative shifting of belt may take place automatically at the end of each stroke, without stopping the machine a belt shifter and its operating lever are provided on the machine.

Trip dogs are mounted at both ends of the table. At the end of each stroke, these dogs strike against the operating lever alternately and the belt shifted accordingly. Thus, table movement is reversed automatically.

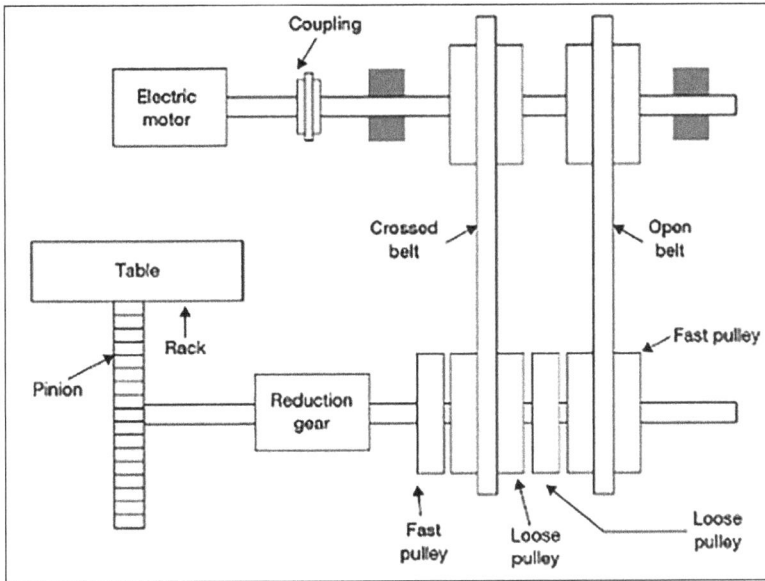

Driving mechanism of planer.

3.5 Different Operations on Lathe

The working of the lathe machine changes with every operation and cut desired. There are a lot of operations used for using the lathe machine. Some of the common lathe operations are:

Facing

This is usually the first step of any lathe operation on the lathe machine. The metal is cut from the end to make it fit in the right angle of the axis and remove the marks.

Tapering

Tapering is to cut the metal to nearly a cone shape with the help of the compound slide. This is something in between the parallel turning and facing off. If one is willing to change the angle then they can adjust the compound slide as they like.

Parallel Turning

This operation is adopted in order to cut the metal parallel to the axis. Parallel turning is done to decrease the diameter of the metal.

Parting

The part is removed so that it faces the ends. For this the parting tool is involved to perform the operation. To make the cut deeper the parting tool is pulled out and transferred to the side or the cut and to prevent the tool from breaking.

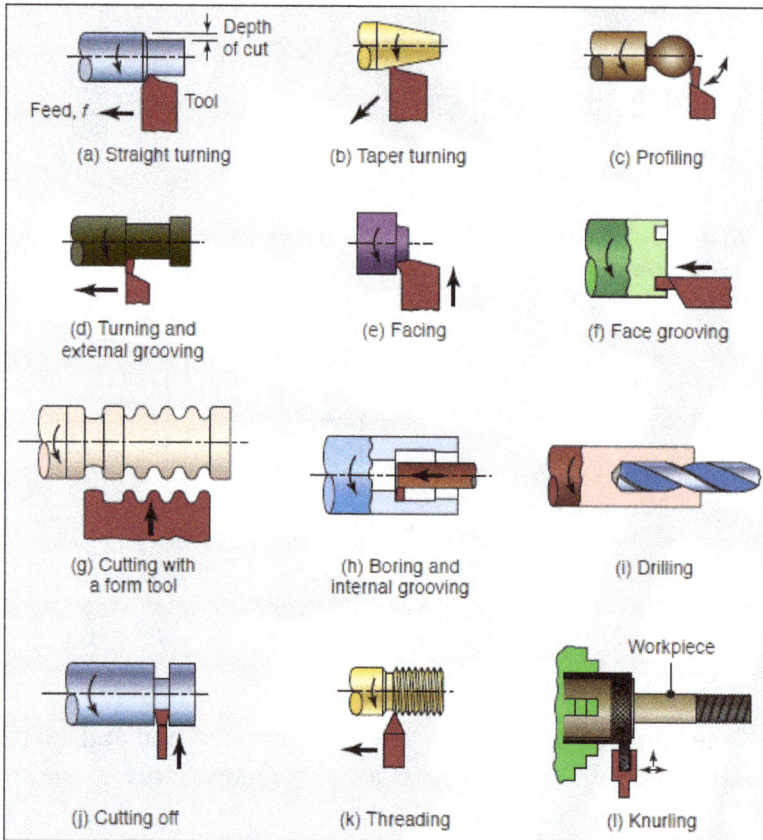

(a) Straight turning   (b) Taper turning   (c) Profiling

(d) Turning and external grooving   (e) Facing   (f) Face grooving

(g) Cutting with a form tool   (h) Boring and internal grooving   (i) Drilling

(j) Cutting off   (k) Threading   (l) Knurling

Shaper Operations

A shaper is a versatile machine tool primarily designed to generate a flat surface by a single point cutting tool. But it may also be used to perform many other operations.

The different operations which a sharper can perform are as follows:

- Machining horizontal surface.

- Machining vertical surface.

- Machining angular surface.

- Cutting slots, grooves and key ways.

- Machining irregular surface.

- Machining splines or cutting gears.

1. Machining Horizontal Surface

A shaper is mostly used to machine a flat, true surface on a work piece held in a vise or other holding devices. After the work is properly held on the table, a shaping tool is set in the tool post with minimum overhand. By the action of reciprocating motion of ram cutting action can be done.

2. Machining Vertical Surface

A vertical cut is made while machining the end of a work piece, squaring up a block or cutting a shoulder.

3. Machining Angular Surface

An angular cut is made at any angle other than a right angle to the horizontal or to the vertical plane.

4. Cutting Slots and Key ways

With suitable tools a shaper can very conveniently machine slots or grooves on a work or cut external key ways on shaft and internal key ways on pulleys or gears.

5. Machining Irregular Surface

A shaper can also produce a contoured surface, i.e., a convex or concave surface or a combination of any of the above surface. To produce a small contoured surface a forming tool is used.

6. Machining Splines or Cutting Gears

By using an index center, a gear or equally spaced spline may be cut.

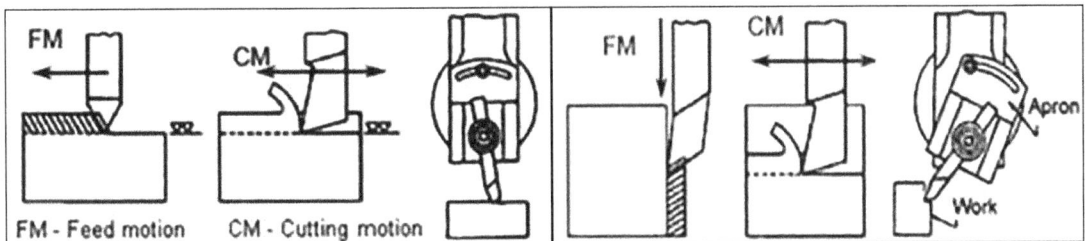

Horizontal surface Vertical surface.

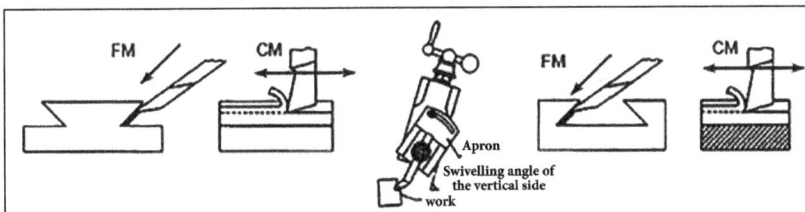

Inclined surfaces (Dovetail slides and guides).Machining of flat surfaces in a shaper.

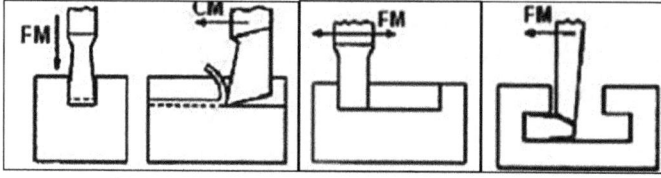

Machining slot Machining pocket Machining T-slot.

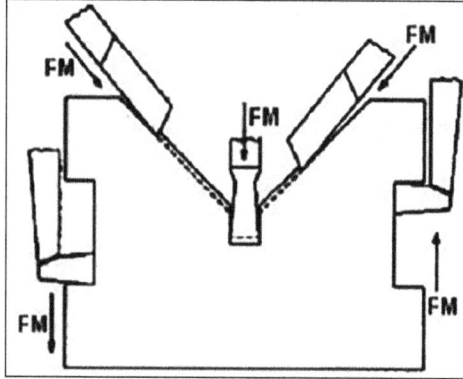

Machining V-block in a shaper.

Planer Operations

The common planer operations performed are as follows:

- Planing horizontal surfaces.

- Planing vertical surfaces.

- Planing curved surfaces.

- Planing slots and grooves.

- Planing at an angle and machining dove-tails.

3.5.1 Simple Problems on Machining Time Calculations

Machining Time

The time required to complete one double stroke, from cutting speed (V_c), is given by,

$$t = \frac{L(1+k)}{1000\,V_c} \text{ min}$$

With a feed of f mm/double stroke, number of double strokes required to machine a surface of width B will be,

$$N_s = \frac{B}{f}$$

Hence, total time for machining the surface will be,

$$t_m = \frac{LB(1+k)}{1000\,V_c f}$$

or, In terms of rem strokes N, the time for machining the surface is given by,

$$t_m = \frac{B}{fN}\,min$$

Machining time (t_m) can also be calculated as follows:

$$t_m = \frac{B}{f}\left(\frac{L}{V_c \times 1000} + \frac{L}{V_r \times 1000}\right)min$$

Where,

B = Width of the job, mm,

f = Feed, ram/stroke,

L = Length of stroke, mm,

V_c = Cutting speed, m/min,

V_r = Return stroke speed, m/min.

If V_{ac} is the average or mean speed, then

$$t_m = \frac{B}{f} \times \frac{2L}{V_{av} \times 1000}\,min$$

$$V_{av} = \frac{2\,V_c \times V_r}{V_c + V_r}\,m/min$$

Turning formulas:

N = Rotational speed of the work piece, rpm.

f = Feed, mm/rev or inch/rev.

v = Feed rate or linear speed of the tool along work piece length, mm/min or inch/min.

$= f\,N$

V = Surface speed of work piece, m/min or ft/min.

$= \pi\,D_o\,N$ (for maximum speed).

$= \pi\,D_{avg}\,N$ (for average speed).

L = Length of cut, mm or inch.

D_o = Original diameter of work piece, mm or inch.

D_f = Final diameter of work piece, mm or inch.

D_{avg} = Average diameter of work piece, mm or in,

$= (D_o + D_f)/2$

d = Depth of cut, mm or in,

$= (D_o + D_f)/2$

t = Cutting time, s or min,

$= L/fN$

MRR = mm³/min or inch³/min,

$= \pi D_{avg} \, d \, f \, N$

Problems

1. Let us calculate the machining time required for machining a surface 450 mm x 600 mm on a shaping machine, using the data given below.

Solution:

Given:

Cutting speed = 7.5 m/min.

Return-to-cutting time ratio = 2:3.

Feed = 2 mm/double stroke.

The clearance at each end = 50 mm.

L_j = 460 mm

B = 600 mm

V_c = 7.5 m/min, k = 2/3

f = 2 mm/double stroke

c = 50 mm

Machining Time t_m

Now, Length of stroke, $L = L_j + 2c$,

$= 450 + 2 \times 50 = 550$ mm

$$t_m = \frac{LB(1+k)}{1000\,V_c f}\,\text{min}$$

Therefore machine time required to machine the job,

$$= \frac{500 \times 600\left(1 + \dfrac{2}{3}\right)}{1000 \times 7.5 \times 2} = 36.67\,\text{min}$$

2. A batch of 800 work pieces is to be produced on a turning machine. Each work piece with length L = 120mm and diameter D = 10mm is to be machined from a raw material of L = 120mm and D = 12mm using a cutting speed V = 32 m/min and a feed rate f = 0.8 mm/rev. Let us determine the times we have to resharpen or regrind the cutting tool.

In the Taylor's expression, use constants as n = 1.25 and C = 175.

Solution:

Given data:

L = 120 mm

D_i = 12 mm

D_f = 10 mm

V = 32 m/min

f = 0.8 mm rev

n = 1.25

C = 175

No of Resharpening Required

From Taylor's tool life equation, we have,

$$VT^n = C$$

$$32 \times (T)^{1.25} = 175$$

$$T = \left(\frac{175}{32}\right)^{1/1.25} = 3.89\,\text{min}$$

Also,

$$V = \frac{\pi DN}{1000}$$

$$32 = \frac{\pi \times 12 \times N}{1000}$$

$$N = 849 \text{ rpm}$$

Now, machining time/piece,

$$= \frac{L}{f \times N} = \frac{120}{0.8 \times 849} = 0.177 \text{ min}$$

Machining time for 800 work pieces = 800 × 0.177 = 141.6 min

Hence, Number of resharpening required,

$$= \frac{141.6}{3.89}$$

= 36.4 or 36 resharpening

3. The shaft shown in the figure below is to be machined on a lathe from a ϕ 25mm bar. Let us calculate the machining time if speed V is 60 m/min, turning feed is 0.2mm/rev, drilling feed is 0.08 mm/rev and knurling feed is 0.3 mm/rev.

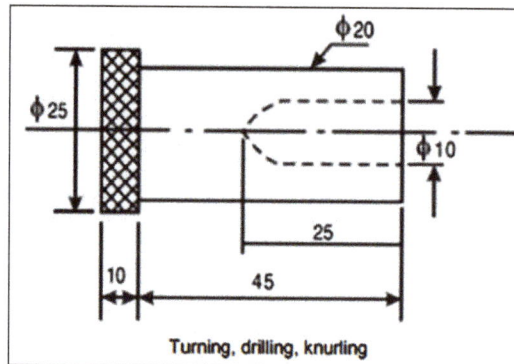

Turning, drilling, knurling

Solution:

Given:

Speed = 60 m/mm

Feed = 0.2mm/rev

Drilling feed is 0.08 mm/rev

Knurling feed is 0.3 mm/rev

To find:

Total machine time.

Step 1

Facing 25 ϕ bar on both ends,

$$N = \frac{1000 \times 60}{25 \times \pi} = 764 \, \text{rpm}$$

Length of cut = 25/2 = 12.5 mm,

$$T_m = \frac{L}{f \times N} = \frac{12.5}{0.2 \times 764} = 0.082 \, \text{min}$$

Time to face on both ends = 2 × 0.082 = 0.164 min

Step 2

Turning ϕ 20mm from ϕ 25mm

$$T_m = \frac{L}{f \times N}$$

$$= \frac{45}{0.2 \times 764} = 0.29 \, \text{min}$$

Step 3

Drilling of 8 mm ϕ hole,

$$N = \frac{1000 \times 60}{10 \times \pi} = 1910 \, \text{rpm}$$

$$T_m = \frac{L}{f \times N} = \frac{25}{00.8 \times 1910} = 0.16 \, \text{min}$$

Step 4

Knurling,

$$N = \frac{1000 \times 60}{\pi \times 25} = 764 \, \text{rpm}$$

$$T_m = \frac{L}{f \times N} = \frac{10}{0.3 \times 764} = 0.04 \, \text{min}$$

Total machining time = 0.164 + 0.29 + 0.16 + 0.04

= 0.65 min

4. A cylindrical stainless steel rod with length L = 150 mm, diameter D_o = 12 mm is being reduced in diameter to D_f = 11 mm by turning on a lathe. The spindle rotates at N = 400 rpm and the tool is traveling at an axial speed of v = 200 mm/min. Let us calculate the following:

- The cutting speed V (maximum and minimum).

- The material removal rate MRR.

- The cutting time t.

- The power required if the unit power is estimated to 4 W.s/mm³.

Solution:

Given:

    L = 150 mm,

    Diameter D_o = 12 mm

    D_f = 11 mm

    N = 400rpm

    U = 200 mm/min

To find:

- The cutting speed V (maximum and minimum).

- The material removal rate MRR.

- The cutting time t.

- The power required if the unit power is estimated to 4 W.s/mm³.

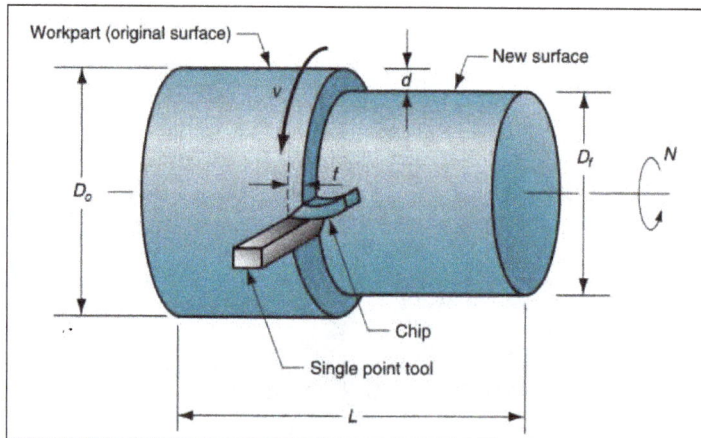

a. The maximum cutting speed is at the outer diameter D_o and is obtained from the expression:

$$V = \pi D_o N$$

Thus,

$$V_{max} = (\pi)(12)\ (400) = 15072 \text{ mm / min}$$

The cutting speed at the inner diameter D_f is given by,

$$V_{min} = (\pi)(11)(400) = 13816 \text{ mm / min}$$

b. From the information given, the depth of cut is:

$$d = (12 - 11)/2 = 0.5 \text{ mm}$$

And the feed is given by,

$$f = u / N$$

$$f = 200 / 400 = 0.5 \text{ mm/rev}$$

Thus the material removal rate is calculated as,

$$\text{MRR} = \pi D_{avg}\ d\ f\ N = (\pi)\ (11.5)\ (0.5)\ (0.5)\ (400)$$

$$= 3611 \text{ mm3 /min}$$

$$= 60.2 \text{ mm3 /s}$$

c. The cutting time is given by,

$$t = 1 / (f.\ N) = (150) / (0.5)\ (400) = 0.75 \text{ min}$$

d. The power required is,

$$\text{Power} = (4)\ (60.2) = 240.8 \text{ W}$$

5. The part shown below will be turned in two machining steps. In the first step a length of $(50 + 50) = 100$ mm will be reduced from Ø100 mm to Ø80 mm and in the second step a length of 50 mm will be reduced from Ø80 mm to Ø60 mm. Let us determine the required total machining time T with the following cutting conditions. Cutting speed V=80 m/min, Feed is f=0.8 mm/rev, Depth of cut = 3 mm per pass.

Solution:

Given:

$$V=80 \text{ m/min}$$

f=0.8 mm/rev

Depth of cut = 3 mm per pass

To find:

Total Machining Time -'T'.

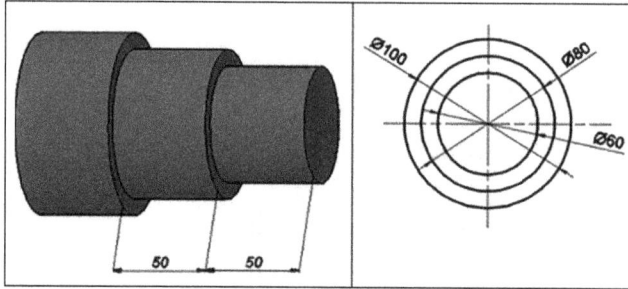

The turning will be done in 2 steps. In first step a length of (50 + 50) = 100 mm will be reduced from Ø100 mm to Ø80 mm and in second step a length of 50 mm will be reduced from Ø80 mm to Ø60 mm.

Step 1

$$N = \frac{1000 \times V}{\pi \times D} = \frac{1000 \times 80}{\pi \times 100} = 255 \text{ rpm}$$

$$\text{Number of passes} = \frac{\text{Depth of material to be removed}}{\text{Depth of cut}}$$

$$= \frac{(100 - 80)}{2 \times 3} = 4$$

$$\text{Time required} = \frac{L}{f \times N} \times 4 = \frac{100}{0.8 \times 255} \times 4 = 1.96 \text{ min}$$

Step 2

To turn from ϕ 80mm to ϕ 60mm for 50mm long,

$$N = \frac{1000 \times V}{\pi \times D} = \frac{1000 \times 80}{\pi \times 80} = 318 \text{ rpm}$$

$$\text{Number of passes} = \frac{(80 - 60)}{2 \times 3} = 4$$

$$\text{Time} = \frac{L}{f \times N} \times 4 = \frac{50}{0.8 \times 318} \times 4 = 0.79 \, \text{min}$$

Total time $= 1.96 + 0.79 = 2.75$ min.

Drilling Machines

4.1 Classification and Constructional Features

In all machining operations, the metal is forcibly ruptured by cutting tools. During the cutting action, the metal comes in contact with the edges of the cutting tool as in drilling and reaming. In drilling, the unstressed grains of the metal are removed continuously by the cutting edge of a drill.

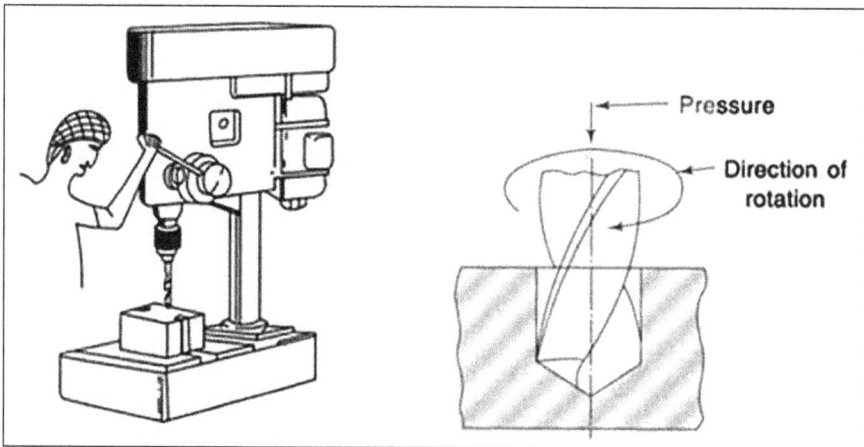

Drilling machine Working of drill.

In a drilling operation, the two cutting lips of the drill work continuously for removing the metal stock. In a workshop, drilling operations are carried out with a large number of machine tools. In order to use the drills efficiently, one must be acquainted with them.

The principal cutting tool used for carrying out drilling operations is the drill. Drilling is the operation of producing fresh round holes with the help of drills. Reaming is the operation of finishing round holes to accuracy with the help of reamers.

Constructional Features of the Drilling Machine

Drilling is a process of making holes of different sizes. A special tool called drill bit is used for drilling. The machine tool used for this is known as drilling machine. The main disadvantage of drilling is that it is not possible to make holes of any odd size. Drill bits are available in standard sizes.

The size that is close to the required size of the hole is to be selected for making the hole. In addition, the surface quality is not very good in drilling. However, drilling is a simple process. The drilling machine also is not very complex. It can be readily used to make holes of different sizes on different jobs. The operator need not be skilled.

4.1.1 Drilling and Related Operations

Operations Performed in Drilling Machine

1. Drilling: It is a process of making round hole by a rotating tool called drill.

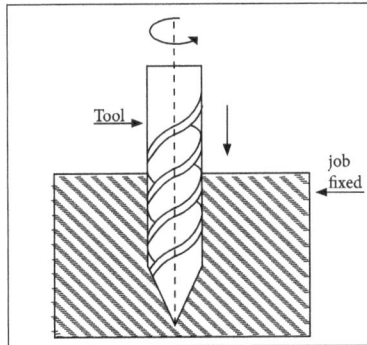

Drilling.

2 Reaming: It is the operation of finishing an existing hole very smoothly and accurately in size.

Reamer is a multi-point rotary cutting tool, generally of cylindrical shape, which removes relatively small amounts of material as it is rotated and fed into a previously drilled or bored cylindrical hole.

It imparts to the hole the necessary smoothness, parallelism, roundness and accuracy in size, It will not correct any error in the hole, because it follows merely the previously drilled hole.

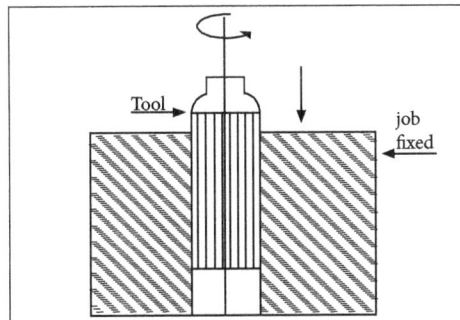

Reaming.

3. Boring: It is the process of enlarging the already existing hole to meet the required size and finish.

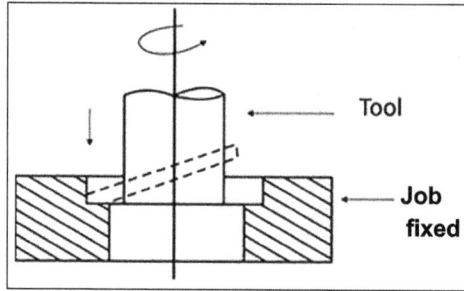

Boring.

4. Counter boring: It is the process of cylindrically enlarging the face of existing hole.

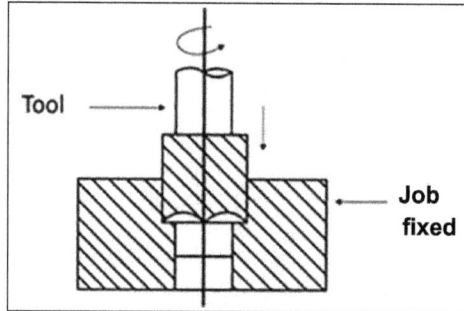

Counter boring.

5. Counter sink: It is the process of enlarging the face of the drilled hole.

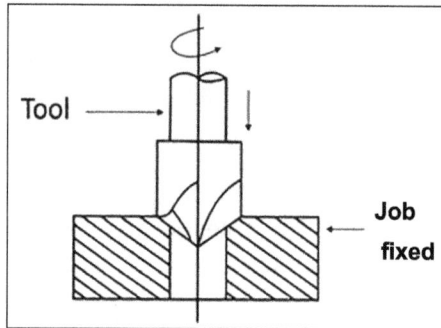

Counter sinking.

6. Spot facing: is the operation of smoothing and squaring the surface around a hole so as to provide a smooth seat for the nut or the head of a screw.

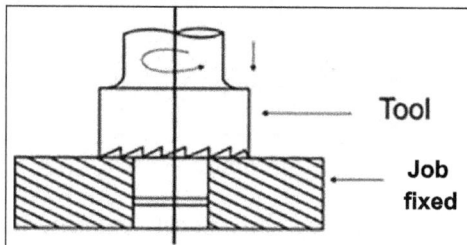

Spot facing.

7. Tapping: It is the process of producing threads in the hole with the help of tool called as tapping.

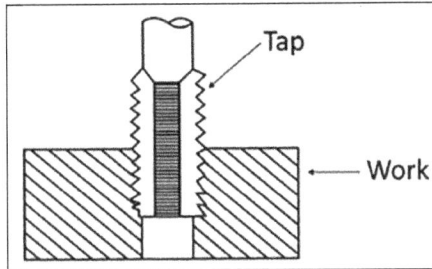

Tapping.

8. Trepanning: It is the process of generating a hole by removing metal along the circumference of a hollow cutting tool. It is used to produce large holes.

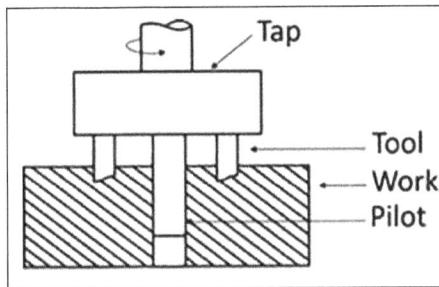

Trepanning.

4.2 Types of Drill

Drilling machines are classified according to the construction and the work formed as follows:

- Portable drilling machine.
- Radial Drilling Machine:
  - Bench type.
  - Floor type.
- Upright drilling machine.
- Sensitive drilling machine.
- Gang Drilling Machine.
- Deep Hole Drilling Machine.

- Automatic Drilling Machine.

- Multi Spindle Drilling Machine.

1. Portable Drilling Machine

The portable drilling machine is a small and compact machine which can be easily moved from place to place by conveniently holding it by hand. It is driven by electric or pneumatic power at high speed. It can drill holes up to 12mm diameter hand drill, ratchet drills and pneumatic arms are examples of portable drilling machines.

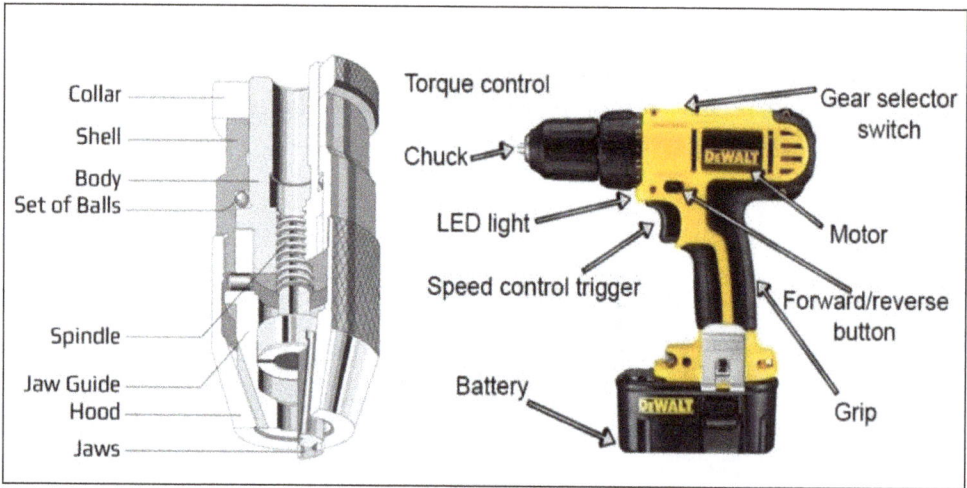

Portable drilling machine.

2. Sensitive Drilling Machine

The sensitive drilling machine is a light, high speed drilling machine. If the machine is mounted on a bench it is called bench type and if mounted on the floor it is called floor type.

This is used generally for light duty and can drill from 1.5mm to 15mm. diameter holes. The drill is fed into the pulley work by hand, only the operator can feel or sense the travel of the drill hence the machine is called as sensitive drilling machine.

- Column: The column is a vertical upright cylinder firmly attached to the base, it supports the table, spindle head, motor and the driving mechanism.

- Table: Table is attached to the column by a clamp. It supports the work piece and the work holding devices, the table can be moved up and can also be rotated around the column.

- Spindle head: The spindle head is mounted at the top of the column, it has a drive motor on one side and spindle assembly on the other side.

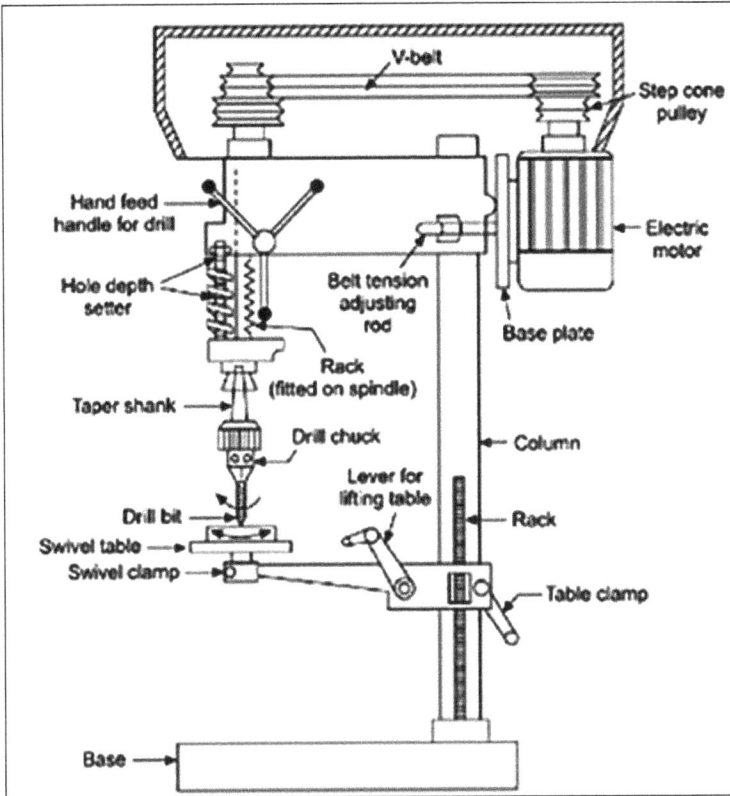

Sensitive drilling machine.

- Drive mechanism: The motor drives the spindle through a V-belt and stepped cone pulley, by shifting the V-belt from one pulley step to another, spindle speeds can be changed.

3. Pillar Type Drilling Machine

This arrangement is mainly for handling medium sized jobs, there are two different constructions.

- Round column known as pillar type drilling machine.

- A box column drilling machine which is known as upright or vertical drilling machine.

This drilling machine is provided with an individual motor drive and the power is transmitted to the spindle through a gear box. The gears are selected with the help of a shift lever. A radial arm is provided to slide over the guide ways on this round vertical column.

Horizontal and vertical movements of the head are achieved by the operation of the gear mechanism provided in the radial arm thus, the arm and table have adjustments to locate the work directly under the drill spindle. Hence this machine is suitable only for light works.

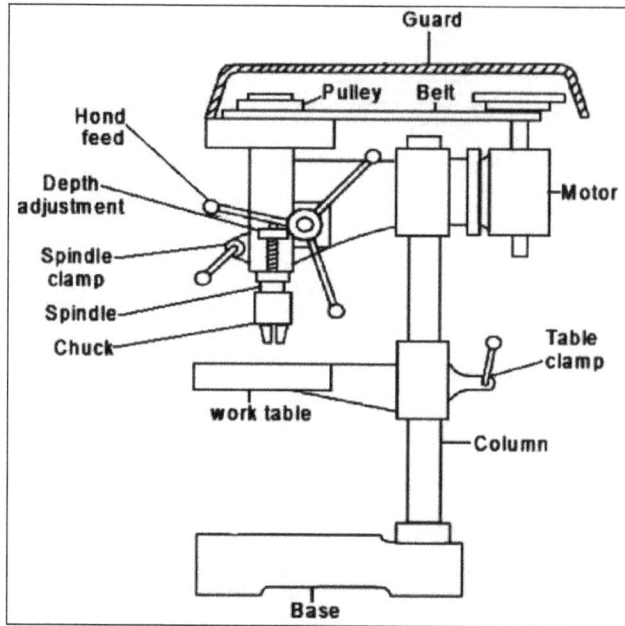

Pillar type drilling machine.

4. Radial Drilling Machine

Radial drilling machine consists of a vertical column with arrangements to raise or lower and to revolve the arm this arrangement helps to accommodate jobs of different heights and to have the radial arm swung around to any position above the machine bed a drill head mechanism is mounted on the radial arm.

This mechanism can be moved horizontally on the guide ways and clamped at any desired position along the arm.

Thus the possible movements are:

- Base: The base is of heavy casting made up of cast iron, it supports the column and other parts of machine.

- Column: The column is a vertical upright cylinder, firmly attached to the base, it supports the table, spindle head, motor and the driving mechanism.

- Table: Table is attached to the column by a clamp. It supports the work piece and the work holding devices. The table can be moved up and can also be rotated around the column.

- Spindle head: The spindle head is mounted at the top of the column. It has a drive motor on one side and spindle assembly on the other side.

- Drive mechanism: A constant speed motor is fixed at the end of the radial arm. This drives a horizontal spindle running along the length of the radial arm.

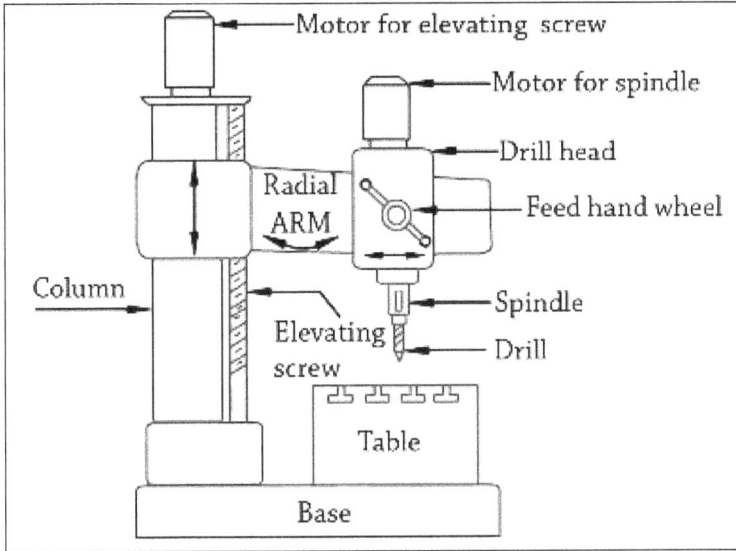

Radial drilling machine.

Advantages of radial drilling machine:

- The arrangement is flexible for any further improvements or alteration.

- The drilling machine is suitable for other operations also like reaming, boring, counter boring, spot facing, tapping, counter sinking, trepanning etc.

- Several holes can be drilled simultaneously by using proper jigs in mass production.

- The operation is quick and cost is less.

- Setting is simple with minimum skill.

5. Gang Drilling Machine

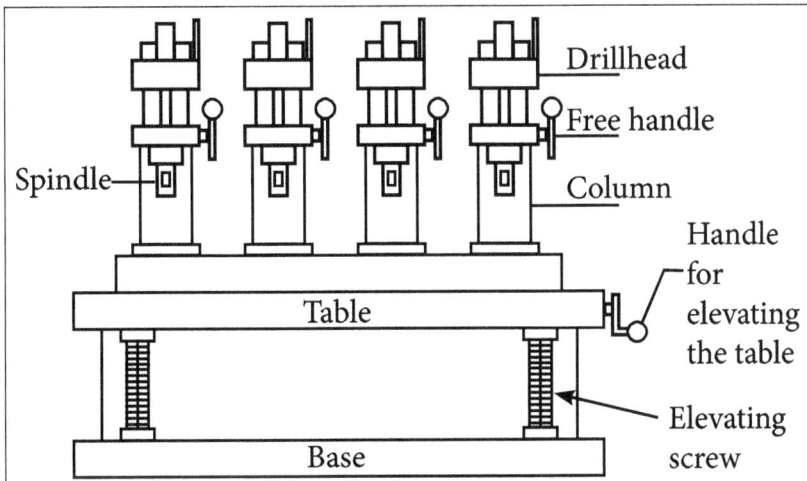

Gang drilling machine.

The machine has a long common base and table, four to six drill heads are mounted on the table. Each head has its own driving motor so that the speeds and feeds of individual units can be controlled independently. Gang drilling machines are used in production line where a series of operation. Like drilling, reaming and tapping are performed on a single job, in a successive manner each spindles performing one particular operation only.

6. Multiple Spindle Drilling Machine

These machines are vertical type machines. They permit drilling of several holes of different diameters simultaneously. Generally the spindles numbering 2 or 3 or even more are driven by only one gear in the head through universal joint linkages. Each spindle is mounted with a twist drill. A jig is used to guide the twist drills.

These machines are mostly used in continuous production shops where several holes of same diameter or different diameters are to be drilled simultaneously and accurately.

Multiple spindle drilling machines.

In mass production, in order to reproduce a pattern of holes in a number of identical jobs, multiple spindle drilling machines is used. This is a special purpose drilling machine and is designed for a particular job or for a particular group of jobs.

7. Deep Hole Drill

Deep hole drilling is required to drill long holes in products such as crank shafts cam shafts, rifle barrels and long shafts.

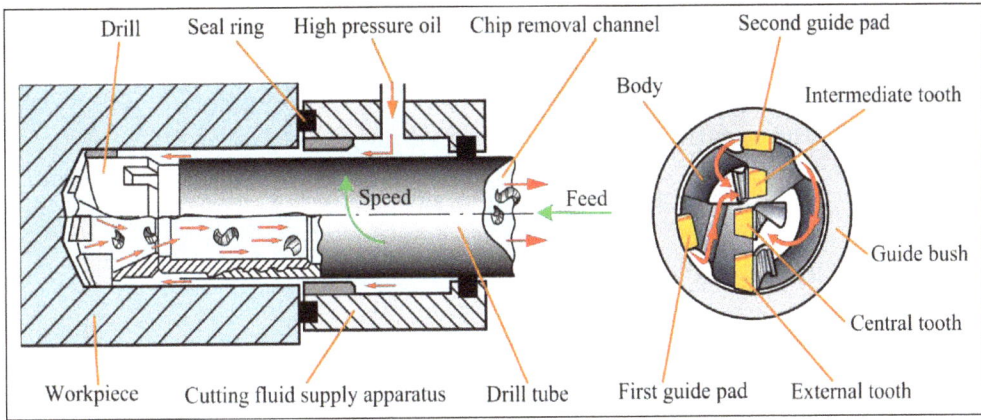

Deep hole drill.

The drill used for drilling deep holes is shown in figure, it has usually one vee flute of 100° included angle, the flute is parallel to the length of drill, the extended lip cuts the metal.

8. Micro Drilling

As the term suggest, micro drilling is the name for the very special world of miniature hole machining, involving dimensions at which many work piece materials no longer exhibit uniformity and homogeneity. Grain borders, inclusion alloy or carbide segregates and microscope voids and problems in micro drilling where holes of 0.02 to 0.0001 inch have been drilled using pivot drills as shown in figure.

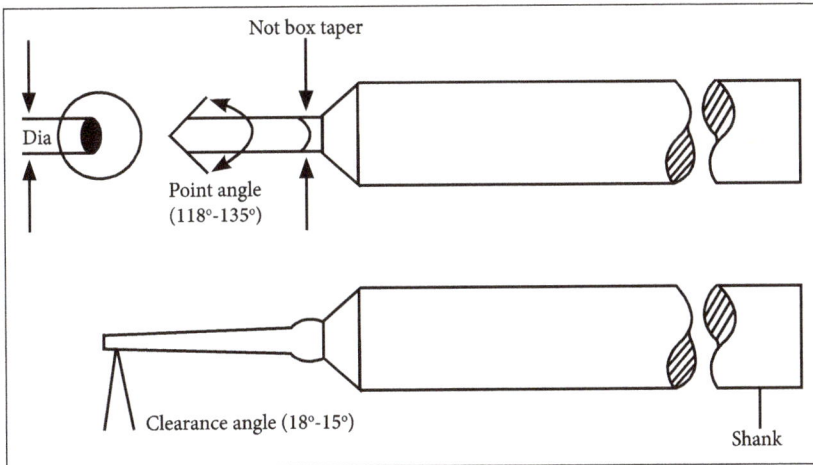

Micro drilling.

Specification of Drilling Machine

- Maximum size of the drill in mm that the machine can operate.

- The maximum dimensions of a job that can mount on a table in square meter.

- Maximum spindle travel in mm.

- Number of spindle speed and range of spindle speeds in rpm.

- Number of automatic spindle feeds or feed range available in mm/ rev

- Morse taper number of the drill spindle nose.

4.2.1 Drill Bit Nomenclature

The various parts and angle of the twist drill are shown in figure:

- Body: It is the part of the drill from its extreme point to the commencement of the neck, if present. Otherwise, it is the part extending up to the commencement of the shank. Helical grooves are cut on the body of the drill.

- Shank: It is the part of the drill by which it is held and driven. It is found just above the body of the drill. The shank may be straight or taper. The shank of the drill can be fitted directly into the spindle or by a tool holding device.

- Chisel Edge: The edge at the end of the web that connects the cutting lips.

- Flutes: Grooves formed in the body of the drill to provide cutting edges, to permit removal of chips and to allow cutting fluid to reach the cutting area.

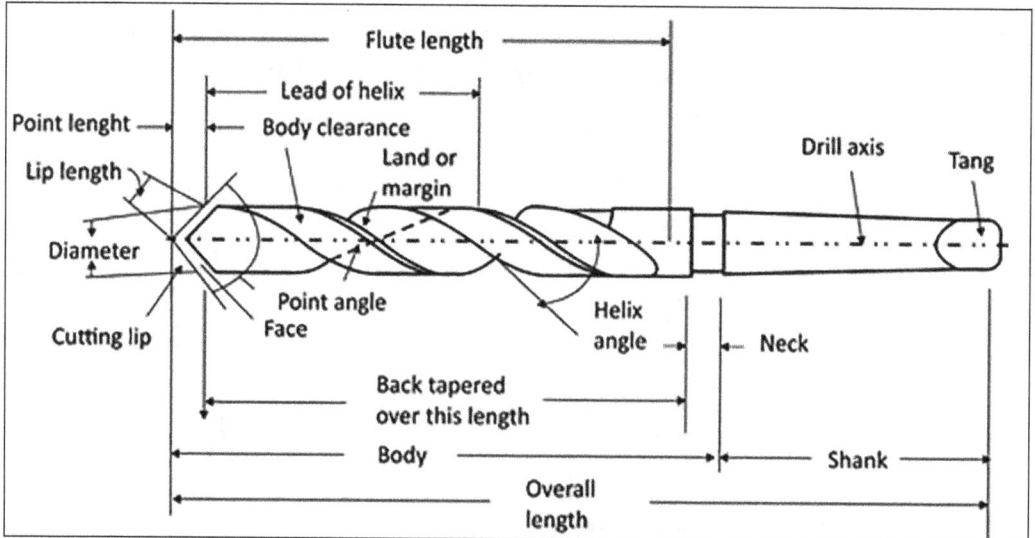

Drilling tool nomenclature.

- Axis: It is the longitudinal centerline of the drill running through the centers of the tang and the chisel edge.

- Lip Relief Angle: The relief angle at the outer corner of the lip.

- Tang: The flattened end of the taper shank is known as tang. It is meant to fit into a slot in the spindle or socket. It ensures a positive drive of the drill.

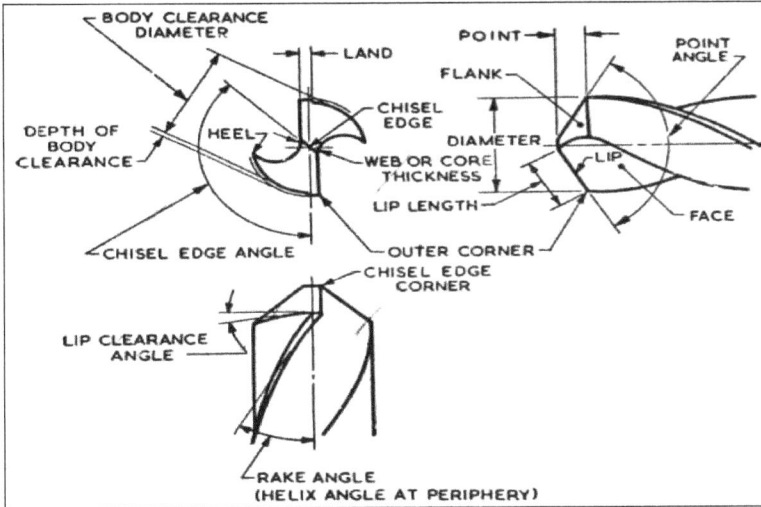

Nomenclature of twist drill.

- Neck: It is the part of the drill, which is diametrically undercut between the body and the shank of the drill. The size of the drill is marked on the neck.

- Point: It is the sharpened end of the drill. It is shaped to produce lips, faces, flanks and chisel edge.

- Lip: It is the edge formed by the intersection of flank and face. There are two lips and both of them should be of equal length. Both lips should be at the same angle of inclination with the axis (59°).

- Helix angle or rake angle: The helix or rake angle is the angle formed by the leading edge of the land with a plane having the axis of the drill. If the flute is straight, parallel to the drill axis, then there would be no rake. If the flute is right handed, then it is positive rake and the rake is negative if it is left handed. The usual value of rake angle is 30° or 45°.

4.2.2 Drill Materials

Carbon Steels

Low and high carbon steels are both used for drill bits, but for different purposes. Soft low carbon steel cannot cut hard metals due to their poor tempers, but they can cut wood. They require sharpening to extend their lifespan. The primary bonus of low carbon steel is its relative inexpensiveness, especially when compared to some more exotic drill bit materials.

High carbon steels have better tempers than low carbon steels, so they require less maintenance, such as sharpening and hold their form and effectiveness longer. They can cut both woods and metals and if available, are preferred to low carbon steels when cutting extremely hard woods.

High Speed Steel

High Speed Steel (HSS) is a special type of carbon steel that is prized for the way it can withstand high temperatures while maintaining structural integrity, specifically its hardness. Friction created by high speed turning can raise temperatures dramatically, but HSS can undergo these types of drillings.

HSS can function at normal temperatures, as well, but only at a level equal to standard carbon steel. HSS can also take coatings, such as titanium nitrate, which give the drill bit better lubricity, decreasing friction and helping to extend the drill bit's life.

Coatings Titanium

Titanium is a corrosion resistant metal and is reasonably light compared to its strength. It is similar to steel in that it has a good fatigue limit and also a high heat limit, although both are less than steels. This longevity makes it attractive for use in repetitive, large runs. It is a very versatile drill bit coating and it can cut a broad variety of surfaces, including many types of steels and irons, as well as wood and plastic.

Zirconium Coating

While not a primary material for drill bits, zirconium coated metals function very well for drill bits. The zirconium nitride coating can increase strength for hard but brittle materials, like steel. The makeup of the zirconium also decreases friction for improved precision drilling.

Cobalt

Cobalt is used for materials that HSS cannot cut, such as stainless steel. It is less susceptible to high temperatures than even HSS so it is not affected by extremely high heat. As a drawback, though, cobalt coatings are excessively brittle.

Cutting Speed

It is the peripheral speed of a point on the surface of the drill in contact with the work piece it is usually expressed in m/min.

Let,

$$V = \pi\, DN\, 1000,\ \text{m/min}$$

Where,

D = Diameter of drill in mm

N = Speed of drill spindle in rpm

V = Cutting speed in m/min

4.3   Introduction to CNC Machines

In order to meet the increasing demand to manufacture complicated components of high accuracy in large quantities, sophisticated technological equipment and machinery have been developed. Production of these components calls for machine tools which can be set up fairly rapidly without much attention.

The design and construction of Computer Numerically Controlled (CNC) machines differs greatly from that of conventional machine tools. This difference arises from the requirements of higher performance levels. The CNC machines often employ the various mechatronics elements that have been developed over the years. However, the quality and reliability of these machines depends on the various machine elements and subsystems of the machines.

There are some of the important constituents parts and aspects of CNC machines to be considered in their designing, for example Machine structure, Guide ways, Feed drives, Spindle and Spindle bearings, Measuring systems, Controls, Software and Operator interface, Gauging, Tool monitoring. The control of a machine tool by means of stored information through the computer is known as Computer Numerically Controlled.

The information stored in the computer can be read by automatic means and converted into electrical signals, which operate the electrically controlled servo systems. Electrically controlled servo systems permits the slides of a machine tool to be driven simultaneously and at the appropriate feeds and direction so that complex shapes can be cut, often with a single operation and without the need to reorient the work piece.

Computer Numerically Control can be applied to milling machines, Lathe machines, Grinding machines, Boring machines, Flame cutters, Drilling machines etc.

Types of CNC Machines

1. According to the type of motion control system:

- Point-to-point position.

- Containing or continuous path.

- Straight out.

2. According to the structure of control system:

- Analog system.

- Digital system.

3. According to the programming mode:

- Incremental programming mode.

- Absolute programming mode.

4. According to the type of control loops:

- Open loop system.

- Closed loop system.

5. According to the main operation:

- CNC lathe.

- CNC milling machine.

- CNC drilling, etc.

Advantages of CNC System over conventional NC system:

- Program Storage: Multiple programs can be stored in the machine by using computer.

- Online Part Programming: The part program can be done online with editing, if required.

- Reliability of System: The data is directly entered with the help of computer, there is no need of punched tape.

- Flexibility of System: The system is too flexible, as new systems can be added at low costs.

- Metric Conversions: Part program, which is written in inches, can be easily converted into millimeter.

Advantage of CNC Machines

- Higher flexibility.

- Increased productivity.

- Consistent quality.

- Reduced scrap rate.

- Reliable operation.

- Reduced non-productive time.

- Reduced manpower.

- Shorter cycle time.

- High accuracy.

- Reduced lead time.

- Just in time (JIT) manufacture.

- Automatic material handling.

- Lesser floor space.

- Increased operation safety.

- Machining of advanced material.

Disadvantages of CNC

- Long-term investment.

- Need to adapt the utility room.

- Ensuring appropriate utilities- compressed air, electricity, etc.

Control System for CNC Machine Tool

1. Open Loop Control Systems

The open-loop CNC systems are always of digital type and are using stepping motor for driving the slides. A stepping motor is defined as one whose output shaft rotates through a fixed angle in response to an input pulse.

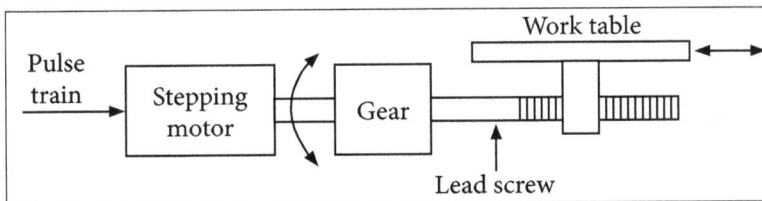

Open loop control system.

The stepping motors are the simplest way for converting digital electrical signals into proportional moment. So, they are a relatively cheap solution to the control problem. Since there is no check on the slide portion, the system accuracy is solely a function of the motor's ability to step through the exact number of steps provided at the input.

2. Closed Loop Control Systems

In closed loop, feedback system is used to close the loop. Position transducer acts as a feedback device, this system measures the axis actual position and compares it with the desired reference position.

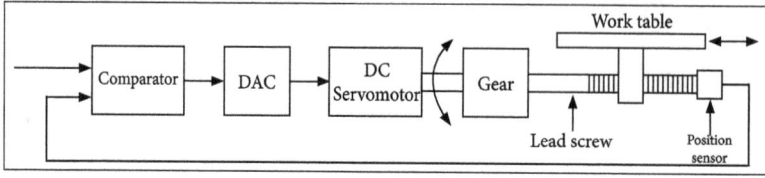

Closed loop control system.

The difference between the actual and the desired values is the error and the control is designed in such a way as to estimate or to reduce the error to a minimum.

3. Adaptive Control

Adaptive Control covers a set of techniques which provide a systematic approach for automatic adjustment of controllers in real time, in order to achieve or to maintain a desired level of control system performance when the parameters of the plant and machine dynamic model are unknown and change in time.

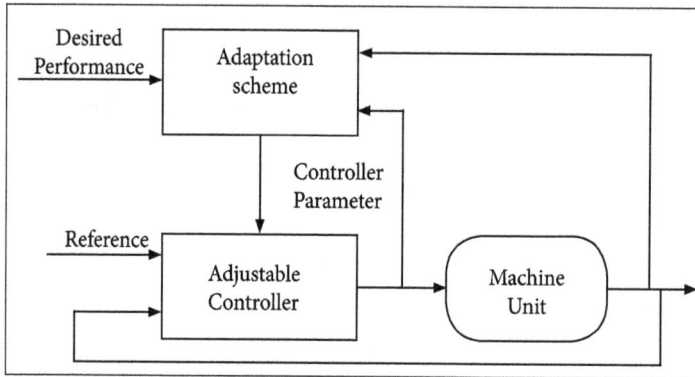

Adaptive Control System.

4.3.1 Principles of Operation

CNC Stands for Computer Numerical Control or Computerized Numerical Control. CNC machine is an Computer Numerical Controlled machine which executes tasks with help of an inbuilt computer which extracts an computer file based on the commands and codes written.

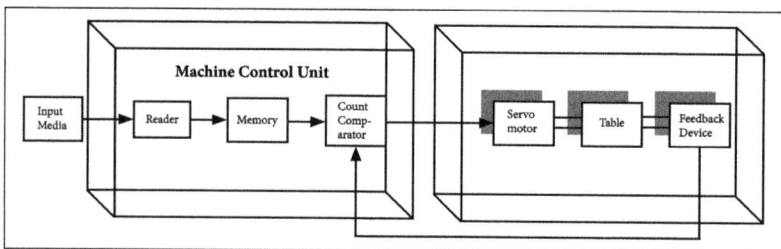

CNC machine working.

Then this command is loaded into post processor which gives electric inputs to the axes drives and spindle which executes the machining requirements. So this is an basic concept of CNC machine working.

So the input is given in the form of G Codes and M Codes thru an Human Machine Interface Device. This commands are read by, an reader and stores into the memory. This memory is in turn converted into Binary codes which gives electric pulses to the Servo drives which actuates the motors and spindle.

Basically tool is fed into an spindle either vertically or horizontally. And the work piece is held into the table by an vice or fixtures. So the rotational motion is thru the spindle and feed motion is thru the machine table. There are three basic axes X, Y, Z.

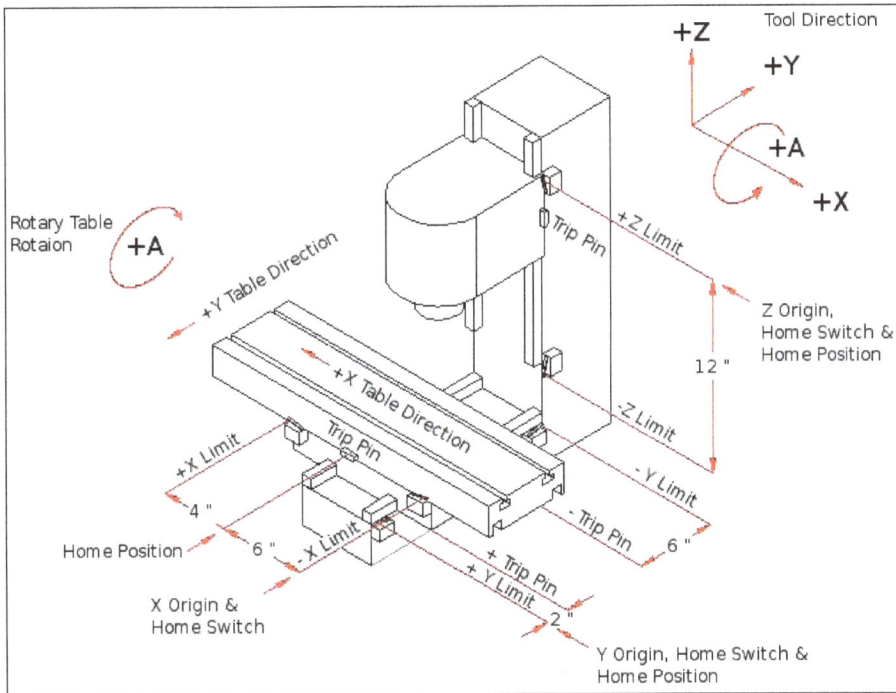

It shows the basic configuration of an CNC machine with its axes. And nowadays all machines consists of an fourth by the use of an rotary table and fifth axes is also added based on requirements. For machining complex parts this fifth axis machines are used.

CNC machines can further categorized into:

- DTC's - Drill tap centres which does the non-accurate operations such as drill and tap and roughing operations.

- Turning centres - CNC lathe machine which is used mostly for rotational jobs.

- Machining centres - Machines which are used for finishing operations. Can be further divided into:

  ◦ Vertical machining centers.

- ○   Horizontal machining centres.

- •   Turn mill center - combined lathes for turning and Milling/ drilling operation.

- •   CNC grinding machines - for grinding operations.

- •   CNC special machines - Controlled by CNC system for special purpose/ specific operations. Can do only Indented operations where as general machines can be used for all operation.

Problems

1 A stepping motor of 200 steps per revolution is mounted on the lead-screw of a NC machine table. The pitch of the screw is 2.5 mm/rev. If the stepping motor receives pulses at a frequency of 2000 Hz, let us derive the linear speed of the table.

Solution:

Given:

Stepper motor speed = 200 steps/rev

Pitch of the screw = 2.5 mm/rev

Stepping motor pulse frequency = 2000 Hz.

Linear Speed of table

$$= \frac{pitch}{step} \times frequency$$

$$= \frac{2.5}{200} \times 2000$$

$$= 25 \, mm/sec.$$

4.4   Axes of NC Machine Coordinate Systems

Controlling a machine tool by means of a set of numerical data is known as numerical control.

The Coordinate System and NC system

In order for the part programmer to plan the sequence of position and movements of the cutting tool relative to the work piece, it is necessary to establish a standard axis system by which relative position can be specified. Using an NC drill press as an example, the drill spindle is fixed in vertical position and the table is moved and controlled relative to the spindle.

However to make things easier for the programmer we adopt the viewpoint that the work piece is stationary while the drill is moved relative to it. Accordingly, the coordinate system of axes is established with respect to the machine table. Two axis are defined as shown in figure.

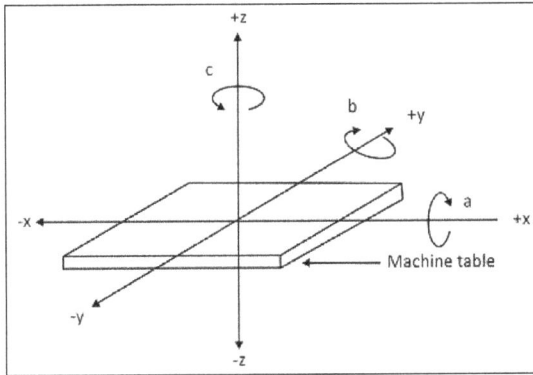

Coordinate system and NC system.

Two axes x and y are defined in the plane of the table. The z axis is defined in the plane perpendicular to the table and the movement in the z direction is controlled by the vertical motion of the spindle. The positive and negative directions of motion of the cutting tools are relative to the table along these axes.

NC drill presses are classified as either two axis or the three axis machines, depending on whether or not they have the capability to control the z-axis.

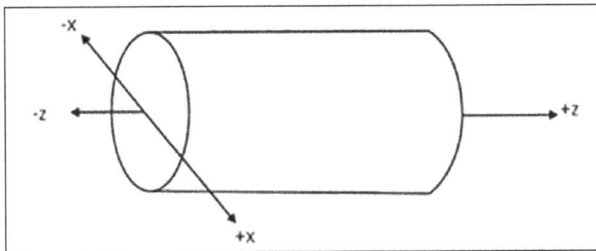

The x and z axes in NC turning.

A numerical control machine and similar machine tools use the axis system similar to the drill system. However in addition to the three linear axes, these machines may possess the capacity to control one or more rotational axes. These axes are used to specify the angle about the x, y and z axes respectively.

To distinguish positive from negative angular motions the right hand rules can be used. Using the right hand with thumb pointing in the direction of positive linear axis direction, the fingers of the hand are curled to point the positive rotational direction.

Main Elements of a NC Machine Tool

1. Part Program

Numerical control is a technique of automatically operating and productive facility, based on a code of letters, numbers and special characters. The complete set of coded instructions; responsible for executing an operation (or a set of operations) is called a part program. This program is translated into electrical signals to drive various motors to operate the machine to carry out the required operations.

Program of instructions: The program of instruction, often called part program is the detailed set of directions for producing a component by the NC machine. Each line of instruction is a mixture of alphabetic codes and numeric data and is punched in a input media (usually paper tape) in a specified format. The input is read by a tape reader which transfers the instructions to a machine controller to operate the machine slides and to generate specific surfaces on the job.

2. Tape Punch

Usually it is a paper tape of 1" width. Paper-muller, aluminum or plastics are also used as tape materials. Paper tapes are cheap and popular but cannot last long. It is treated to resist oil and water.

Mylar tapes are expensive but durable. Mylar tapes are still used by machine manufacturers to store information's as executive tapes. Punching machine of various types are used to key in program instructions to tapes. Presently tapes are prepared by micro-computers by keying in the information from the manuscript.

Once the entire program has been entered, it is checked and corrected if needed and then the computer activates the tap punching unit to produce the tap. The computer can also generate the program print-out through its printer.

3. Tape Reader

A tape reader reads the hole pattern on the tape and converts the patterns to a corresponding electrical signal.

4. Machine Controller

Controller receives the electrical signals from tape reader or an operating panel and causes NC machine to respond. Figure shows the function of a NC controller.

It contains a decoder/encoder, an interpolator and facilities to execute auxiliary functions which are machine dependent. The decoder/encoder receives the data and stores them in two separate memory locations.

One for the part geometry data and the other form the process data. Process data includes switching functions for adjusting feed rates, spindle speeds, tool changes, cutting fluid applications etc. Geometric data consists information about tool motions, tool length, tool radius, tool compensation etc.

As the machine is to shape complex surfaces at a constant feed rate, signals must be given to various slides and spindles so that the individual motions can be integrated to produce the required shape which can be represented by complex curve or simple lines.

The interpolator breaks down these curves into small individual increments for each controlled motion of the machine tool. Controller also interfaces various machine units like drive motors, transducers and other control functions of the machine tools.

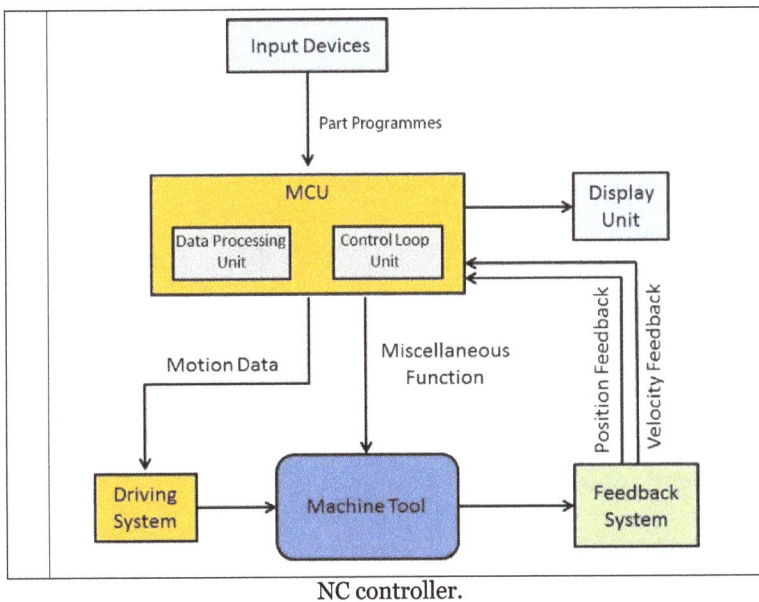

NC controller.

5. NC Machine

NC machine responds to the electrical signals from the controller. Accordingly the machine executes various slide motions and spindle rotations to manufacture a part. Any NC machine can be considered as a general purpose machine tool fitted with drive motors and other auxiliary functions of the machine.

It consists as usual the work table, spindle and other hard wares as a general purpose machine contains. Transducers are fitted to feedback data on the positions of the slide ways, for the r.p.m. of the spindle and for the amount of cut on the job.

NC machine tools range from single spindle drilling machine to complex machines having multiple motions, tool changers, high capacity tool magazines and multi-axis control.

4.4.1 Basics of Manual Part Programming Methods

The program for machining any type of work piece varies from machine to machine. Various controllers use different syntax during instructing the machine tools. It is necessary that, the programmer understands the different processes involved by carefully studying the drawing, fixtures and machine tools. A typical block diagram of this process is shown in Figure.

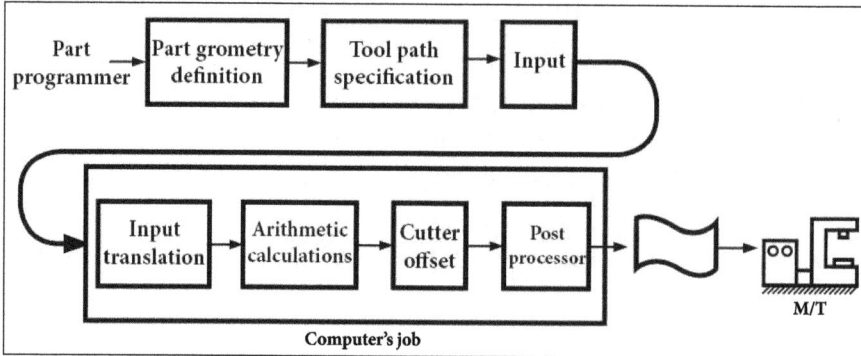

Manual part programming.

Canned Cycle in Manual Part Programming

A fixed cycle is a combination of machine moves resulting in a particular machining function such as drilling, milling, boring and tapping. By programming one cycle code number, as many as distinct movements may occur.

These movements would take blocks of programmer made without using fixed or canned cycles. The corresponding instructions of a fixed cycle are already stored in the system memory.

Manual Part Programming

The programmer first prepares the program manuscript in a standard format.

Manuscripts are typed with a device known as flexo writer, which is also used to type the program instructions. After the program is typed, the punched tape is prepared on the flexo writer. Complex shaped components require tedious calculations.

This type of programming is carried out for simple machining parts produced on point-to-point machine tool.

To be able to create a part program manually, need the following information:

- Knowledge about various manufacturing processes and machines.
- Sequence of operations to be performed for a given component.
- Knowledge of the selection of cutting parameters.

- Editing the part program according to the design changes.

- Knowledge about the codes and functions used in part programs

The advantages of writing a part programme with these structures are:

- Reduced lengths of part programme.

- Less time required for developing the programme.

- Easy to locate the fault in the part programme.

- No need to write the same instructions again and again in the programme.

- Less memory required in the control unit.

The program is a set of an instruction for machine tool which defines the tool position relative to the work piece.

Each line of program is called as block, which consists of an operation number word, data word etc.

The format of each block is as follows:

N....G....X....Y....Z....U....V....W....F....S....T....M;

- N is a sequence number and used for giving the number to the lines. It is written as N001, N002, N003,..... This will identify the block number.

- G is a preparatory function which changes the control mode of the machine and is called as G-codes. G codes are followed by two digit number. It is written as G00, G01, G02, G03, ..., etc.

- X, Y, Z and U, V, W represents co-ordinate position of tools. For two axes system, only two letters are specified. In case of multiple axes other additional letters, i.e., U, V, W are specified. X, Y, Z can be positive or negative according to dimension.

- F is the feed rate function, which defines feed rate of operation. For example F100, it means feed rate is 100 mm/min. If it is specified once, then no need to specify again. It continues unless and until another value is specified.

- S is a cutting speed function which specifies spindle speed.

- T is a Tool change function. Generally, all the CNC machines are having ATC for programming; each tool is associated with an index number.

- M is a miscellaneous function which is generally called as M-codes. By specifying M codes, other auxiliary operations are performed.

- End of Block (EOB) is written after each and every line.

Problems

1. A 110 mm long cylindrical rod of ϕ 75 mm is to be turned into a component as shown in figure, using a CNC lathe. Let us write a CNC program for manufacturing this component.

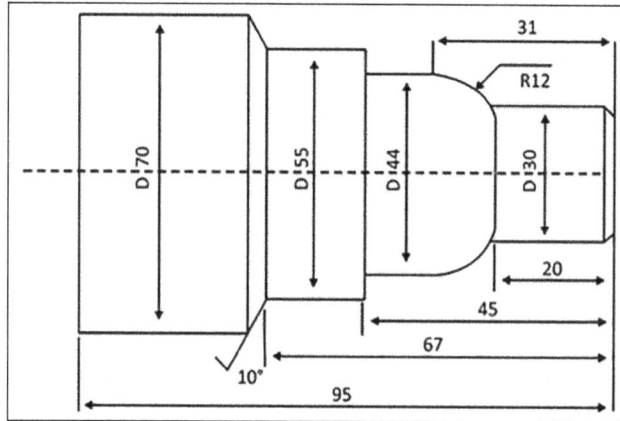

(All dimensions are in mm).

CNC Program:

    059X0Z125

    G27

    N1 T0101M03

    G96 V180

    G00 VI80

    G01 Z74 Z0 M08

    G01 X30

    G01 X-1.6

    G00 X70Z2

    G71 P50Q60 10.5K0.1 D4F0.35

    G26

    N3 T0404 M03

    G96 V250

    N50G46

    G00Z26 Z1F0.15

G01X30 Z-1F0.2

G03X44 Z-30.909 R12 F0. 15

G01Z-45F0.2

G0lX55R1.5

G0lZ-67

G0lX7A105 C-0.4

G00X70.5

G40

N60

G26

M09

M30

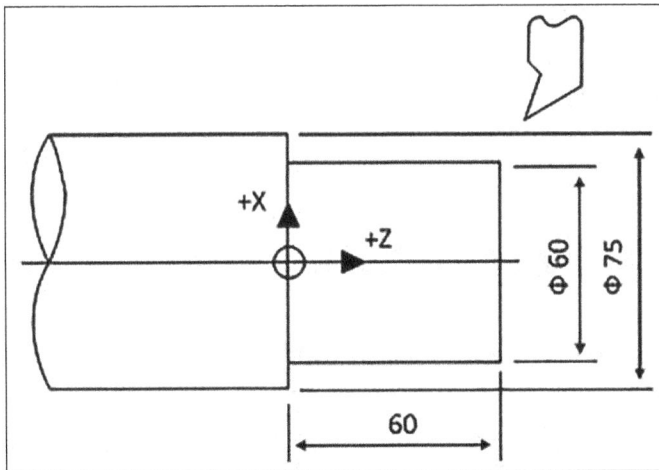

2. Let us write a manual part program to turn the component shown on a CNC Lathe from 75 mm bar stock. The following data may be assumed:

- There will be two rough turnings and one finish turning. The first cut is with a depth of 3 mm for a length of 58 mm, the second with a depth of 3 mm for a length of 59 mm and the third with a depth of 1.5 mm for the full length of 60 mm.

- The shoulder of the work-piece is also machined during each cut.

- The spindle speed is 400 rpm and the feed rate is 0.5 mm/ rev.

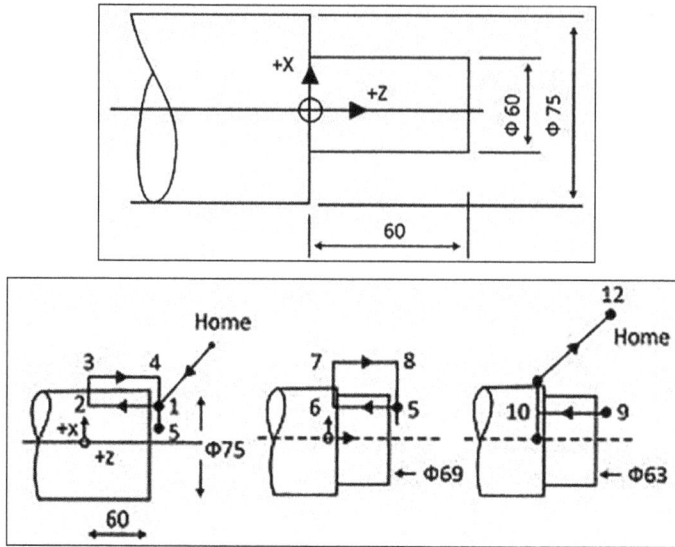

First rough turning Second rough turning Finish turning.

DOC = 3mm; DOC = 3mm; Length =59mm DOC = 1.5 mm.

Length = 58 mm Length = 60mm.

[BILLET × 75 ZB0.

N00 G90 G71...Absolute programming mode and data input is in Metric mode.

N01 G95 G28 U0 W0...Feed in mm/rev and go to home position.

N02 M06 T0101...Tool change Tool no. 1 (Roughing tool).

N03 S400 M03...Spindle speed 350 rpm and clockwise −direction.

N04 G00 X69 Z61 M08...Rapid positioning at point 1 and coolant ON.

N05 G01 Z2 F0.5...First rough turning up to point and feed 0.5 mm/rev.

N06 X77...Positioning at point-3.

N07 G00 Z61...Move to point-4 at rapid speed.

N08 X63...Positioning at point 5 at rapid speed.

N09 G01 Z1 f0. 1...Second rough turning up to point 6.

N10 X77...Positioning at point 7.

N11 G00 Z61...Rapid positioning at point 8.

N12..X60...Rapid positioning at point 9.

N13 G01 Z0 f0.5...Finish turning up to point 10.

N14 X77...Positioning of tool at point 11.

N15 G00 G28 U0 W0 M09 ... Rapid traverse to home position and coolant OFF.

N16 M05 M0...Spindle stop and program stop.

3. The bracket shown in figure is 15 mm thick. Its profile is slightly over sized by about 1 mm. Let us write a APT program to do finish milling of the profile of the bracket. The following data may be assumed:

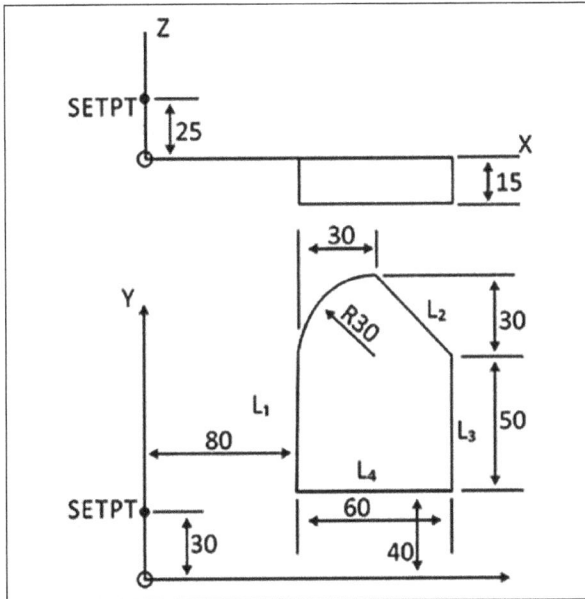

- A 20 mm end-mill cutter is to be used.

- The X, Y, Z axes are as shown in the figure.

- The start point is at (0, 30, 25).

- For milling, the spindle speed is 1740 rpm and the feed rate is 500 mm/min.

- The post processor statement is MACHIN/UNI. (16)

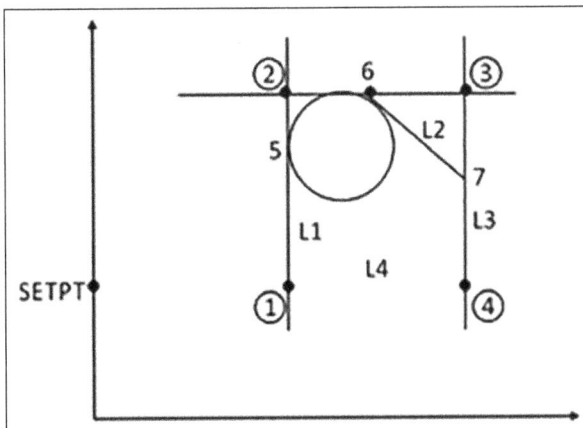

Solution:

Part no/6:

SP = POINT/0, 30,-15

PT1 = POINT/80, 40, -15

PT2 = POINT/80, 120, -15

PT3 = POINT/140, 120, -15

PT4 = POINT/140, 40, - 15

PT5 = POINT/80, 90,-15

PT6 = POINT/110, 120, -15

PT7 = POINT/140, 90, -15

L1 = LINE/PT1, PT2

L2 = LINE/PT6, PT7

L3 = LINE/PT4, PT7L4 = LINE/PT1, PT4

L5 = LINE/PT2, PT3

C1 = CIRCLE/PT5, PT6, RADSUS, 30

INTOL/0.01

CUTTER/20

SPINDL/1740

FEDRAT/MMPM, 500

TL LFT/OH

FROM SP

GOTO/PT1

GOLFT/L1 PAST L5

GORGT/L5 PAST L3

GOFWD/L3 PAST L4

GORGT/L4 PAST L1

GOLFT/L1 TANTO C1

GORGT/C1 PAST L2

GORGT/L2 PAST L3

GOFWD/L3 PAST L4

GORGT/L4 PAST L1

GOTO/SP

COOLNT/OFF

SPINDL/OFF

MACHIN/UH2

Milling Machines

5.1 Classification and Constructional Features

Milling machine is one of the important machining operations. In this operation the work piece is fed against a rotating cylindrical tool. The rotating tool consists of multiple cutting edges (multi-point cutting tool). Normally axis of rotation of feed given to the work piece.

Milling operation is distinguished from other machining operations on the basis of orientation between the tool axis and the feed direction, however, in other operations like drilling, turning, etc. the tool is fed in the direction parallel to axis of rotation.

The cutting tool used in milling operation is called milling cutter, which consists of multiple edges called teeth. The machine tool that performs the milling operations by producing required relative motion between work piece and tool is called milling machine. It provides the required relative motion under very controlled conditions.

Normally, the milling operation creates plane surfaces. Other geometries can also be created by milling machine. Milling operation is considered an interrupted cutting operation teeth of milling cutter enter and exit the work during each revolution. This interrupted cutting action subjects the teeth to a cycle of impact force and thermal shock on every rotation.

Classification

Milling machines are classified in a variety of ways. According to the drive, milling machines are classified as:

- Cone-pulley belt drive.
- Individual motor drive.

According to design, milling machines are classified as:

- Column and knee-type milling machine.
- Planer milling machine.
- Fixed bed-type milling machine.
- Special milling machines. Such as rotary table, duplicating and profiling.

Depending upon the position of the spindle, milling machines are classified as:

- Horizontal spindle milling machines.

- Vertical spindle milling machines.

The spindle of the horizontal milling machine is horizontal to the worktable, while the spindle of the vertical milling machine is at right angles to the worktable. In a vertical milling machine, the cutter can be raised or lowered by an adjustment of the spindle head. In all milling machines, the worktable can be moved to any position to carry out the operations.

Machine size Milling machines 'are specified by the longitudinal travel of the worktable, the horse power of the main motor, the type of milling machine and its model.

Construction of Milling Machine

Base

The base of the machine is a gray iron casting accurately machined on its top and bottom surface and serves as a foundation member for all the other parts which rest upon it. In some machines, the base is hollow and serves as a reservoir for cutting fluid.

Column

The column is the main supporting frame mounted vertically on the base. The column is box shaped, heavily ribbed inside and houses all the driving mechanisms for the spindle and table feed. The front vertical face of the column is accurately machined and is provided with dovetail guide ways for supporting the knee. The top of the column is finished to hold an over arm that extends the outward at the front of the machine. The below figure shows the various parts of Milling Machine:

- Base

- Elevating screw

- Knee

- Knee elevating handle

- Cross feed handle

- Saddle

- Table

- Front brace

- Arbor support

- Arbor

- Overhanging arm

- Milling cutter

- Column

- Cone pulley

- Telescopic feed shaft

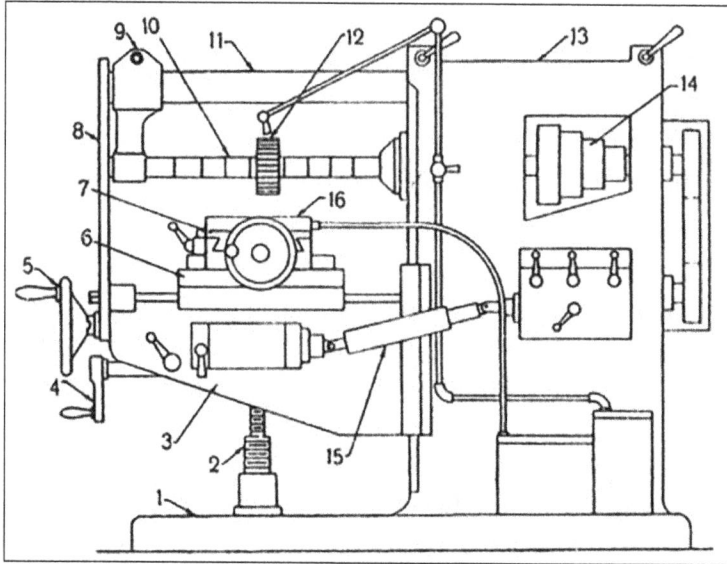

Knee

The knee is a rigid grey iron casting that slides up and down on the vertical ways of the column face. The adjustment of height is effected by an elevating screw, mounted on the base that also supports the knee. The knee houses the feed mechanism of the table. The top face of the knee forms a slide way for the saddle to provide cross travel of the table.

Saddle

On the top of the knee is placed the saddle, which slides on guide ways set exactly at 90° to the column face. A cross feed screw near the top of the knee engages a nut on the bottom of the saddle to move it horizontally, by hand or power. The top of the saddle is accurately machined to provide guide ways for the table.

Table

The table rests on base on the saddle and travel longitudinally. The top of the table is accurately finished and T-slots are provided for clamping the work and other fixtures on it. A lead screw under the table engages a nut on the saddle to move the table horizontally by hand or by power.

Over Hanging Arm

The overhanging arm that is mounted on the top of the column extends beyond the column face and serves as a bearing support for the other end of the arbor. The ram is adjustable so that the bearing support may be provided nearest to the cutter.

Front Brace

The front brace is an extra support that is fitted between the knee and the over arm to ensure further rigidity to the arbor and the knee. The front brace is slotted to allow for the adjustment of the height of the knee relative to the over arm.

Spindle

The spindle of the machine is located in the upper part of the column and receives power from the motor through belts, gears and clutches and transmit it to the arbor. The front end of the spindle just projects, from the column face and is provided with a tapered hole into which various cutting tools and arbors may be inserted.

Arbor

An arbor may be considered as an extension of the machine spindle on which milling cutters are securely mounted and rotated. The arbors are made with taper shanks for proper alignment with the machine spindles having taper holes at their nose. The arbor may be supported at the farthest end from the overhanging arm or may be of cantilever type which is called stub arbor.

Universal Milling Machine

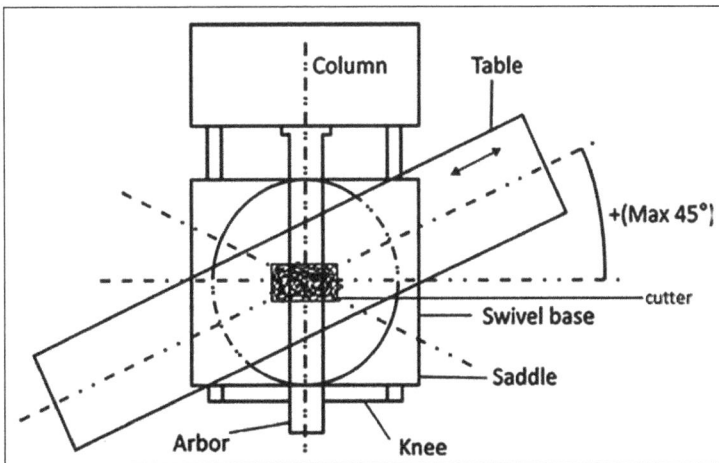

Universal milling machine.

The machine is similar to the horizontal milling machine in all respects with an additional swiveling movement for the table which rests on a graduated swivel base, the table can be rotated about the vertical axis through 45° of the axis an either side for helical milling operations the table is turned to the required angle and fed. Special attachments like, vertical milling attachment, rotary table attachment etc. are used in universal milling machines.

Omniversal Milling Machine

In this machine, the knee is attached to a circular base so that it can be swiveled about a horizontal axis parallel to the spindle.

The table has all the four movements of a universal milling machine (i.e., longitudinal movement, cross movement. vertical movement and rotation about vertical axis) as well as a fifth movement, rotation about a horizontal axis parallel to the spindle. Using this additional movement, tapered spiral grooves can be machined (e.g., reamers. bevel gears). This machine is mainly used in tool rooms.

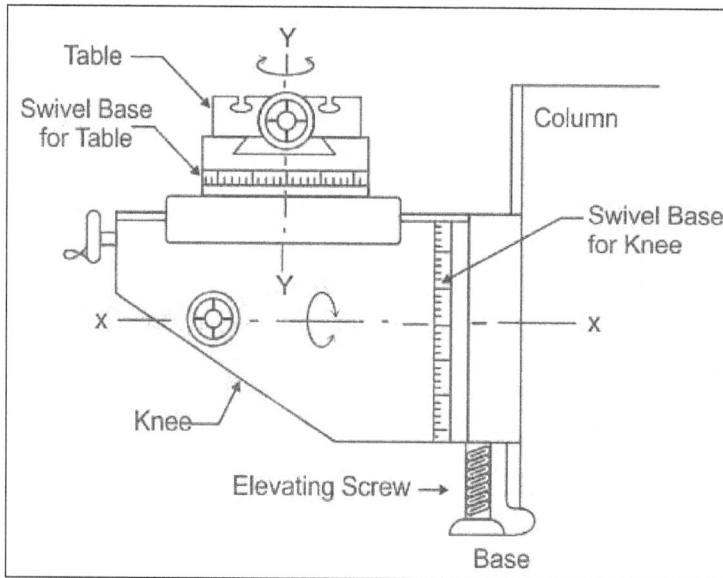

Omniversal milling machine.

Vertical Milling Machine

- Base: The base is the foundation of the milling machine. It supports the entire structure of the machine such as the column and table. It gives strength and rigidity to the machine. The base of some milling machines are hollow and are used as reservoir for the cutting fluid.

- Column: Column is the main supporting frame mounted vertically on the base. The column houses the motor, other driving mechanisms and the spindle. The front vertical face of the column is provided with guide ways for supporting the knee.

- Saddle: The saddle supports and carries the table it slides on the guide ways on the top of the knee. These guide ways are exactly at 90° to the face of the column.

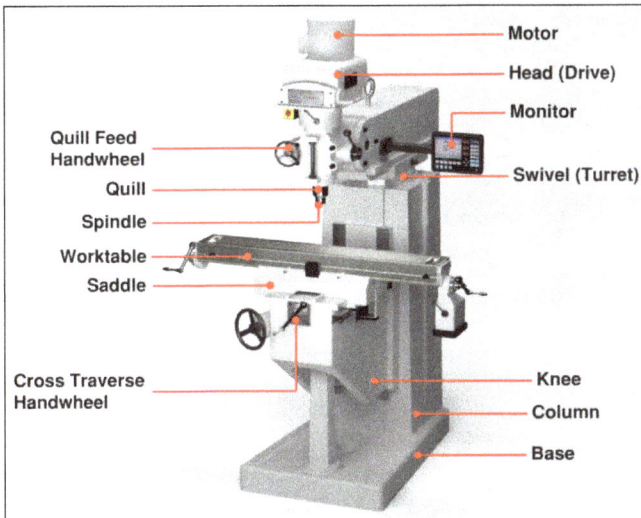
Vertical milling machine.

- Knee: The portion which projects from the column and is made to slide up and down is known as knee it supports the saddle and the table the provision of the knee can be adjusted by an elevating screw, provided at the bottom of the knee.

- Table: The table is placed on the guide ways on the saddle and it travels longitudinally, the top face of the table has T-slots, which are used to clamp the work piece firmly.

- Spindle: The spindle of the machine is located in the upper part of the column it obtains power from the electric motor through belts and gears.

5.1.1 Milling Cutters Nomenclature

As far as metal cutting action is concerned, the pertinent angles on the tooth are those that define the configuration of the cutting edge, the orientation of the tooth face and the relief to prevent rubbing on the land.

The terms defined below and illustrated in the figures, are important and fundamental to milling cutter configuration.

- Outside Diameter: The outside diameter of a milling cutter is the diameter of a circle passing through the peripheral cutting edges. It is the dimension used in conjunction with the spindle speed to find the cutting speed (SFPM).

- Root Diameter: This diameter is measured on a circle passing through the bottom of the fillets of the teeth.

- Tooth Face: The tooth face is the surface of the tooth between the fillet and the cutting edge, where the chip slides during its formation.

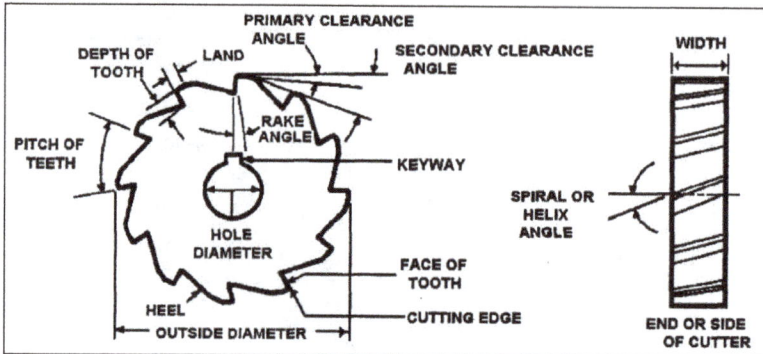

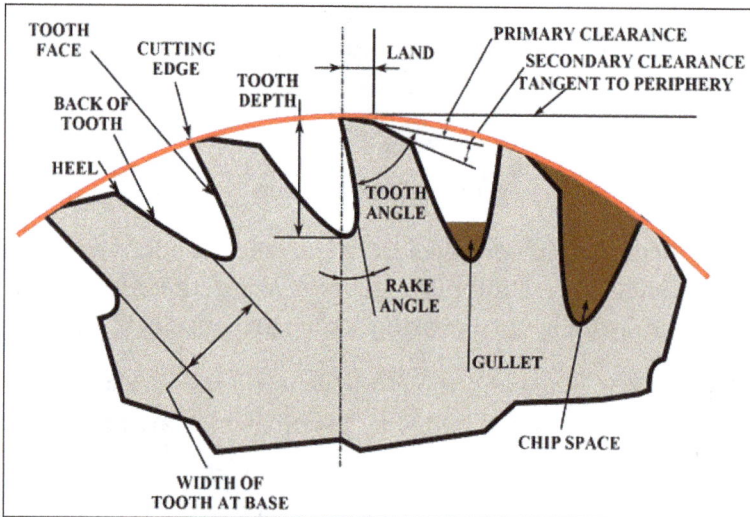

Milling cutter nomenclature.

- Tooth: The tooth is the part of the cutter starting at the body and ending with the peripheral cutting edge. Replaceable teeth are also called inserts.

- Land: The area behind the cutting edge on the tooth that is relieved to avoid interference is called the land.

- Flute: The flute is the space provided for chip flow between the teeth.

- Fillet: The fillet is the radius at the bottom of the flute, provided to allow chip flow and chip curling.

The terms defined above apply primarily to milling cutters, particularly to plain milling cutters. In defining the configuration of the teeth on the cutter, the following terms are important:

- Gash Angle: The gash angle is measured between the tooth face and the back of the tooth immediately ahead.

- Peripheral Cutting Edge: The cutting edge aligned principally in the direction of the cutter axis is called the peripheral cutting edge. In peripheral milling, it is this edge that removes the metal.

- Face Cutting Edge: The face cutting edge is the metal removing edge aligned primarily in a radial direction.

In side milling and face milling, this edge actually forms the new surface, although the peripheral cutting edge may still be removing most of the metal. It corresponds to the end cutting edge on single point tools:

- Radial Rake Angle: The radial rake angle is the angle between the tooth face and a cutter radius, measured in a plane normal to the cutter axis.

- Relief Angle: This angle is measured between the land and a tangent to the cutting edge at the periphery.

- Clearance Angle: The clearance angle is provided to make room for chips, thus forming the flute. Normally two clearance angles are provided to maintain the strength of the tooth and still provide sufficient chip space.

- Axial Rake Angle: The axial rake angle is measured between the peripheral cutting edge and the axis of the cutter, when looking radially at the point of intersection.

- Blade Setting Angle: When a slot is provided in the cutter body for a blade, the angle between the base of the slot and the cutter axis is called the blade setting angle.

- Plain milling cutter.

- Side milling cutter.

- Slitting saw.

- Single angle and double angle milling cutter.

- End mill.

- Face milling cutter.

- T-slot milling cutter.

- Fly cutter.

- Form cutter.

It is cylindrical in shape and has cutting teeth on the periphery. The teeth may be straight or helical. This cutter is used to produce flat surfaces parallel to the axis of the spindle.

This cutter is used to produce T-slots for machine tool tables and fixtures. The cutter is similar to milling cutter but it has a tapered shank.

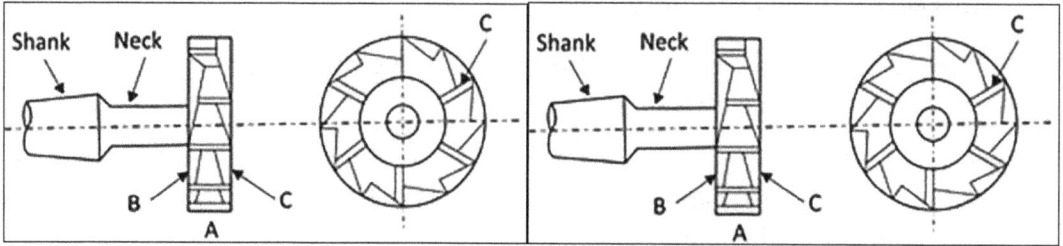

T - slot milling cutter Involute gear milling cutter.

It consists of a single point tool with the cutting edge formed to produce the desired contour in the work-piece form cutter. It has teeth curved outward on the periphery and, when used, produces a concave semi-circular surface on the work piece.

5.2   Milling Operations

The following are the different operations performed in a milling machine:

- Plain milling cutter
- Side milling cutter
- Slitting saw
- Single angle and double angle milling cutter
- End mill
- Face milling cutter
- T-slot milling cutter
- Fly cutter
- Form cutter

Plain Milling

The plain milling is the operation of production of a plain, flat, horizontal surface parallel to the axis of rotation of a plain milling cutter. The operation is also called slab mining. To perform the operation, the work and the cutter are secured properly on the machine.

The depth of cut is adjusted by rotating the vertical feed screw of the table and the machine is started after selecting proper speed and feed. The plain milling operation is illustrated in figure.

Slab Milling Cutter

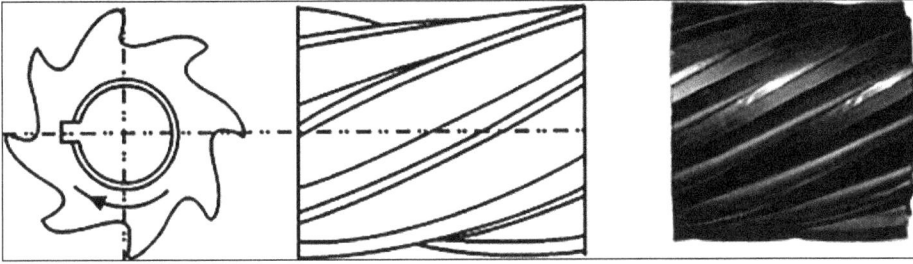

Slab or plain milling cutter.

In this cutter, the width is greater than the diameter, the helix angle is large and the cutter has a less number of teeth. These cutters nave nicked teeth.

Face Milling

The face milling operation is performed by a face milling cutter rotated about an axis perpendicular to the work surface. The operation is carried in a plain milling machine and the cutter is mounted on a stub arbor to produce a flat surface. The depth of cut is adjusted by rotating the cross feed screw of the table. The face milling operation is illustrated in figure.

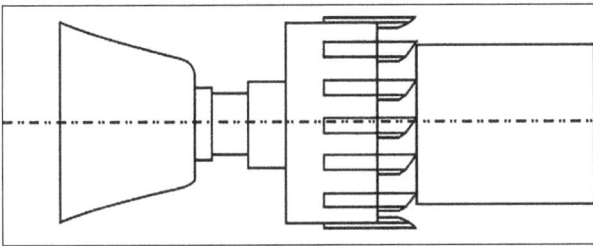

Face milling operation

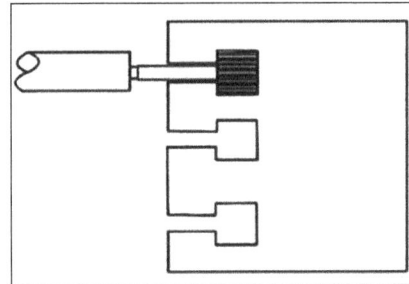

T-slot milling operation.

Side Milling Cutter

A side milling cutter has cutting edges on the periphery as well as on its sides. In a half-side milling cutter, the cutting edges are on the periphery and on one side.

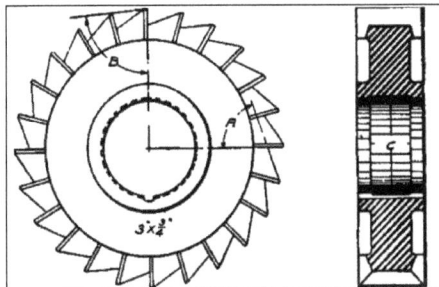

Side milling cutter.

Slitting Saw

In this cutter, teeth are on the periphery only. Cutter width varies from 0.8 to 5mm.the sides are round concave to prevent them from rubbing the work piece.

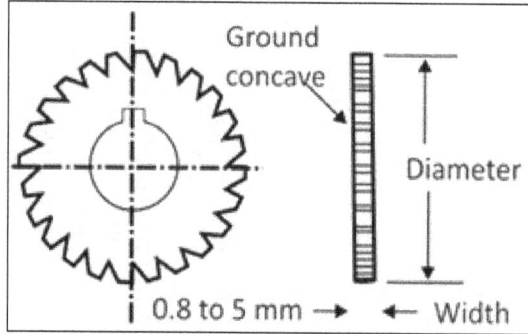

Slitting saw

Angle Milling Cutter

These are used to machine angular surfaces. Since angle milling cutter has cutting teeth on the conical surface as well as on the large flat face, these cutters are designated by the included angle between the conical side and flat face.

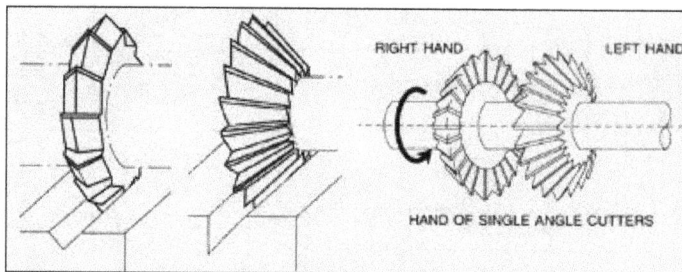

Angle milling cutter.

End Mills

In the end mills the cutting teeth are on the periphery as well as on the end face. The peripheral teeth may be straight or helical commonly used in vertical milling machine.

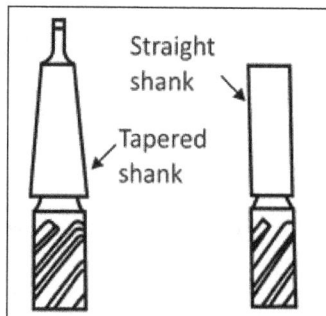

End mill.

Face Milling Cutter

It has cutting teeth inserted on the cutter body. These teeth project a little outside the body so that the cutter end has cutting edges.

Slot Milling Cutter

(a) T - slot milling cutter.

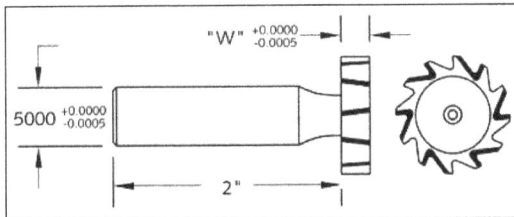

(b) Involute gear milling cutter.

This cutter is used to produce T-slots for machine tool tables and fixtures. The cutter is similar to milling cutter but it has a tapered shank.

Fly Cutter

It consists of a single point tool with the cutting edge formed to produce the desired contour in the work-piece form cutter. It has teeth curved outward on the periphery and, when used, produces a concave semi-circular surface on the work piece.

Operations Performed in Milling Machine

Milling machines are mostly general purpose machine tools and used for piece or small lot production.

In general, all milling operations can be grouped into two types. They are:

- Peripheral milling.
- Face milling.

Peripheral Milling

Here, the finished surface is parallel to the axis of rotation of the cutter and is machined by cutter teeth on the periphery of the cutter. Below figure schematically shows the peripheral milling operation.

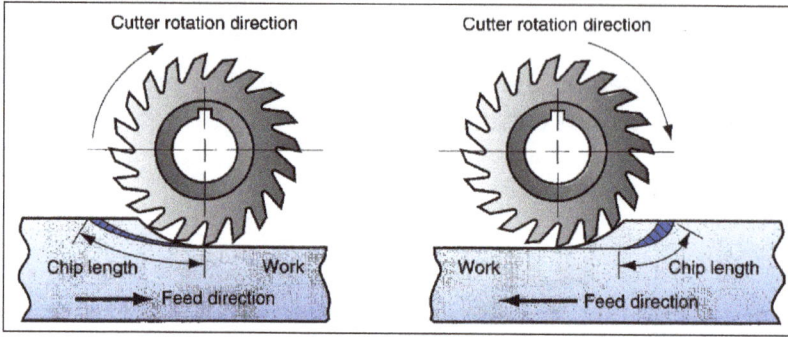

Schematic view of the peripheral milling operation.

Face Milling

Here, the finished surface is perpendicular to the axis of rotation of the cutter and is machined by cutter teeth on the periphery and the flat end of the cutter. The peripheral cutting edges do the actual cutting, whereas the face cutting edges finish up the work surface by removing a very small amount of material. Below figure schematically shows the face milling operation.

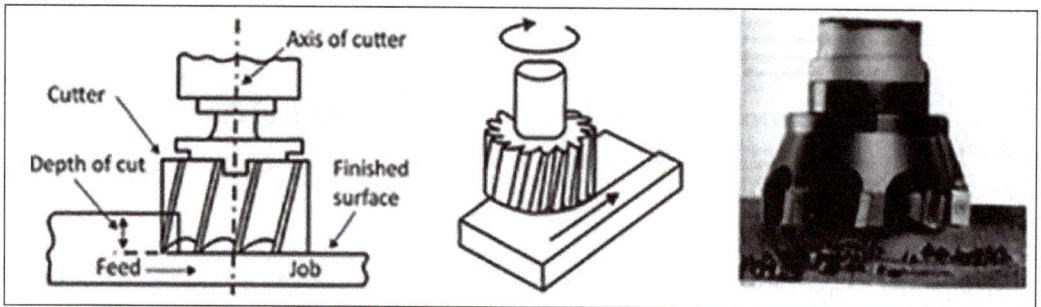

Schematic view of the face milling operation.

Special Type - End Milling

It may be considered as the combination of peripheral and face milling operation. The cutter has teeth both on the end face and on the periphery. The cutting characteristics may be of peripheral or face milling type according to the cutter surface used. Below figure schematically shows the different end milling operation.

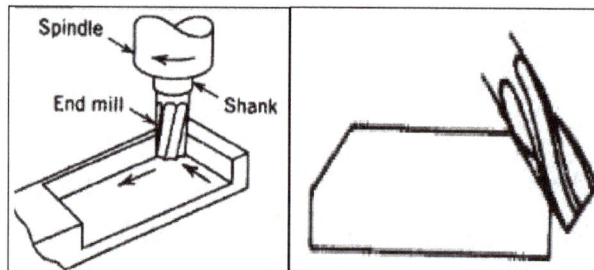

Face milling by end mill and Angular milling by end mill.

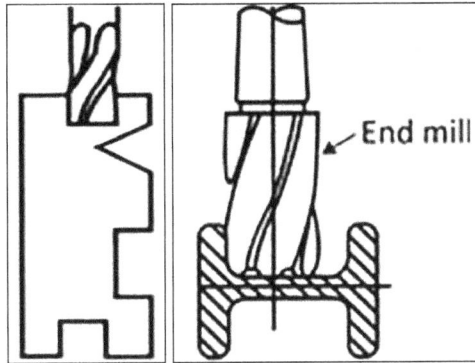

Slotting by end mill End milling using a corner radius. Schematic views of the different end milling operations.

According to the relative movement between the tool and the work, the peripheral milling operation is classified into two types. They are: up milling and down milling.

Climb Milling and its Advantages

Down milling, which is also called climb milling, is the process of removing metal by a cutter which is rotated in the same direction of travel of the work piece.

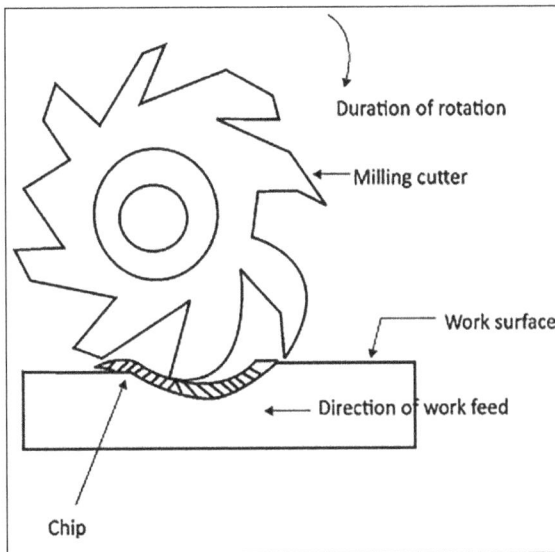

Climb Milling.

Advantages

- The fixture design becomes easier as the direction of the cutting force is such that it tends to seat the work firmly in the work holding devices.

- The chips are also disposed of easily and do not interfere with the cutting.

- The coolant can be poured directly at the cutting zone where cutting force is maximum.

Difference between Horizontal and Vertical Milling Machines

S. No.	Horizontal Milling Machine	Vertical Milling Machine
1.	Milling cutter rotates about horizontal axis.	Cutter rotates about vertical axis.
2.	Swindle can't be moved up and down.	Spindle head can be moved up and down.
3.	Spindle arbor cannot swiveled about horizontal axis.	Spindle head may or may not swiveled about horizontal axis.
4.	Base and column are separate castings.	Base and column are integral castings.
5.	Milling cutter is mounted at the middle of spindle or arbor.	Milling cutter is mounted at at the end of spindle.
6.	Work table can be swiveled.	Work table may or may not be swiveled.
7.	It is suitable for external surface operation.	It is suitable for internal surface operation.

5.2.1 Up Milling And Down Milling Concepts

According to the relative movement between the tool and the work, the peripheral milling operation is classified into two types. They are:

    (i) Up milling   (ii) Down milling

Up Milling or Conventional Milling

Here, the cutter rotates in the opposite direction to the work table movement. In this, the chip starts as zero thickness and gradually increases to the maximum. The cutting force is directed upwards and this tends to lift the work piece from the work holding device. Each tooth slides across a minute distance on the work surface before it begins to cut, producing a wavy surface.

This tends to dull the cutting edge and consequently have a lower tool life. As the cutter progresses, the chip accumulate at the cutting zone and carried over with the teeth which spoils the work surface. Figure (a) schematically shows the up milling or conventional milling process.

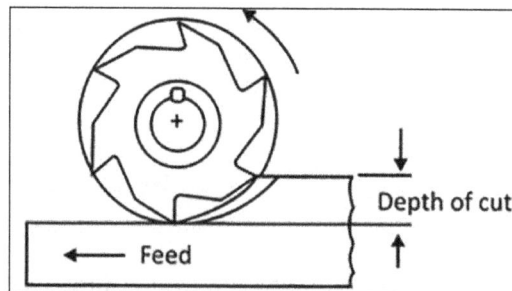

(a) Conventional or Up milling.

Schematic views of (a) Up milling process and (b) Down milling process.

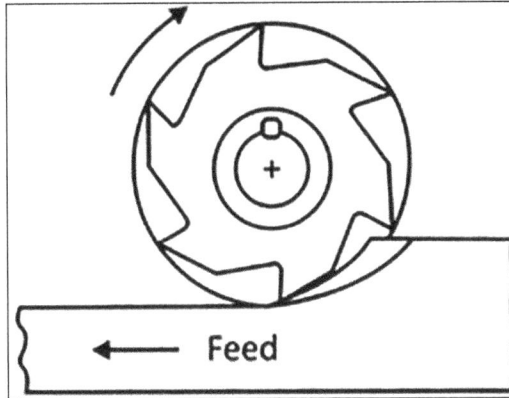

(b) Climb or Down milling.

Down Milling or Climb Milling

The cutter rotates in the same direction as that of the work table movement. In this, the chip starts as maximum thickness and gradually decreases to zero thickness. This is suitable for obtaining fine finish on the work surface. The cutting force acts downwards and this tends to seat the work piece firmly in the work holding device.

The chips are deposited behind the cutter and do not interfere with the cutting. Climb milling allows greater feeds per tooth and longer tool life between regrinds than up milling. Figure (b) schematically shows the down or climb milling process.

Difference between up Milling and Down Milling

Up Milling	Down Milling
The work piece is fed opposite to the cutter rotation.	The work piece is in the same direction of the cutter rotation.
The chip thickness is minimum at the beginning of the cut.	The chip thickness is maximum at the beginning of the cutting.
Poor surface finish.	Better surface finish.

5.2.2 Various Milling Operations

1. Single-piece Milling

One work piece is held at a time in the vice or fixture and fed through the milling cutter. The work piece is removed, the next one is placed in the fixture, the mill table is returned to the starting point and then the milling operation is carried out on the next work piece.

2. String Line Milling

A string milling process is shown in the figure (b) ,In this process, a series of identical work pieces are mounted in line parallel to the table and fed in consecutive order against the milling cutter. In order to minimize idle time between the two pieces, the work pieces should be held as close together as possible.

3. Reciprocal Milling

As shown in the figure (c), fixtures are mounted on each end of the worktable and the milling cutter is held in the center. When the left hand work piece is under milling operation, the operator unloads the right hand work piece and vice versa.

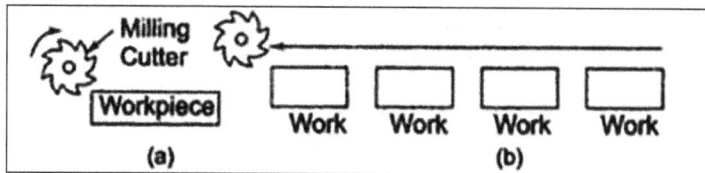

(a) Single-piece milling (b) String line milling.

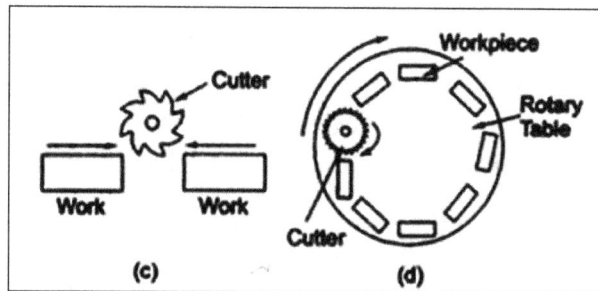

(c) Reciprocal milling (d) Rotary milling.

4. Progressive Milling

In progressive milling, two or more operations are performed simultaneously on identical pieces. Two different types of cutters are mounted on arbors. In this system it is necessary to move a partially finished work piece to the next station in the future and insert the next in its place.

5. Index Milling

In this process, duplicate fixtures are mounted on the table. The table is so pivoted that it can be rotated and indexed into position. The advantage of index milling is that the operator can load and unload the work piece while the operation is going on.

6. Rotary Milling

In this system, a number of fixtures are mounted on a large rotary table. The table rotates the work piece consecutively under the cutter. It is a quick mass production process, since the operator can remove and insert the new work piece without stopping the table.

Milling fixture are generally classified as follows:

1. According to type of milling operation performed on the work:

- Face milling fixture.
- Slap milling fixture.
- Slotting fixtures.
- Straddle milling fixtures.

2. According to the method of clamping the work piece:

- Hand clamping fixtures.
- Power clamping fixtures.
- Toggle fixtures.

3. According to the method of location of the work piece:

- Center fixture V-block fixtures.
- Stud fixtures.

4. According to the method of presenting the work piece to the cutter:

- Cradle fixtures.
- Rotary fixtures.
- Indexing fixtures.
- Rise and fall fixture.
- Progressive fixtures.

A recent development in the design of magnetic chucks is the development of magnetic milling chucks. Vacuum chucks are used for holding non-ferrous and non-magnetic parts for milling operations.

Plain Milling Fixtures

Plain milling designed fixtures are used for holding components of complicated shape. The operation is carried out with plain milling cutters while the work is held in the fixture. The fixture consists of a special adjustment of jack screws or spring loaded rest pins for each component. The major drawback of these fixtures are their inability to hold more than one kind of component.

Straddle Milling Fixtures

A fixture designed for milling two sides simultaneously is known as a straddle milling fixture. The operation is performed by mounting two cutters on the milling machine arbor, as shown in the figure below. The fixture consists of a cast iron base provided with a stop pin or rest pin. The rest pin is provided for locating the work. Usually, in a straddle milling fixture, several components are held at a time, which helps in rapid production.

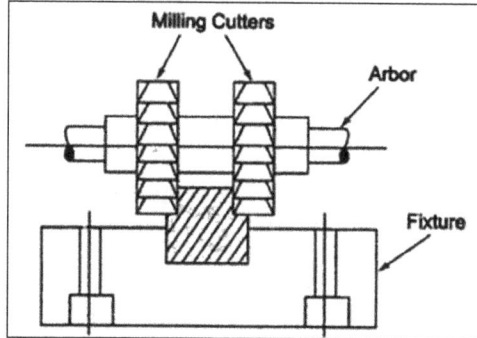

Straddle milling with the help of milling.

Keyway Milling Fixtures

Keyway cutting with the help of milling fixtures is carried out on horizontal milling machines. The cutter is fixed rigidly on the arbor and the shaft is held in the fixture. A commonly used fixture is a vice with special vice jaws, as shown in the figure. An open-type keyway can be milled on a vertical milling machine with the help of end mill cutters.

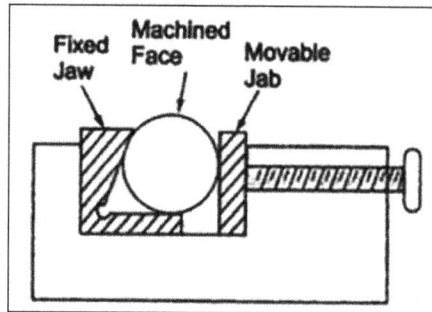

Keyway milling.

5.3   Indexing: Simple and Compound

The process of dividing a circular or straight part into equal spices by means of a dividing head is known as indexing. The indexing head is also known as dividing head. The indexing head is an attachment that forms a part of milling machines, by means of which the circumference of a cylindrical part can be divided into any number of equal spaces.

It is also used for imparting a rotary motion to the work. For example, if some circular part requires 24 equally spaced grooves, the dividing head is used to rotate the work 1/24 after cutting each groove.

The three systems of indexing used on a milling machine are:

- Simple indexing

- Compound indexing

- Differential indexing

Simple Indexing

Simple indexing on a milling machine is carried out by using either a plain indexing head or a universal dividing head. This method of indexing involves the use of crank, worm, worm wheel and index head. The worm wheel generally carries 40 teeth and the worm is single threaded. With this arrangement, when a crank completes one revolution, the work wheel rotates through 1/40th of a revolution.

Similarly, a worm wheel rotates through 2/40 (1/20)th of a revolution and so on. Thus, for one revolution of the work-piece, a crank needs to make 40 revolutions. The holes in the index plate further help in sub-dividing the rotation of the work piece.

Suppose the work is to be divided into a number of parts. The corresponding crank movement will be as follows.

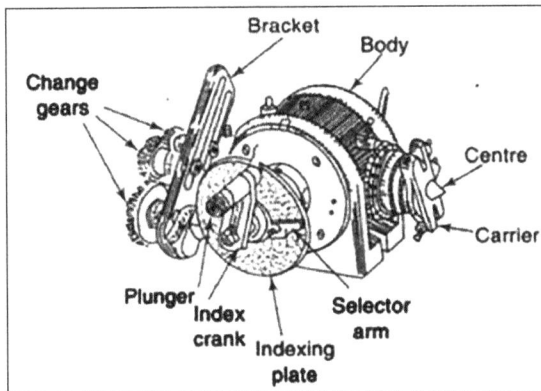

Universal dividing head.

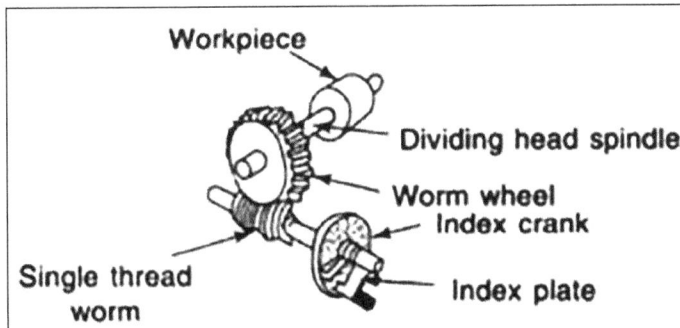

Simple indexing.

For dividing the work in two equal parts a crank will make:

- For each division, (40/20) =20 Revolution.

- For 5 division, (40/5) = 8 Resolution.

- For 8 division, (40/8) = 5 Resolutions.

- For 29 division, (40/29) = 1(11/29) Resolutions.

In the last example above. 40/29 is not a whole number. This indicates that the crank moves by I rotation and 11/29 parts of the second revolution. In the fractional system, the numerator shows the number of holes to be moved and the denominator shows the number of holes on the index plate to be used.

Thus, in this indexing system, for each division on the job, the crank will move through one revolution and II holes on the 29 holes index circle on the index plate.

Compound Indexing

When the number of divisions required on the job is outside the range of simple indexing, the method of compound indexing is used. The operation is carried out by providing two separate simple indexing movements:

- By turning the crank in the same way as in simple indexing.

- Again turning the index plate and the crank either in the same or in the opposite directions.

The principle of compound indexing can be best understood from the following practical example.

Suppose the crank is turned 3 holes in a 15-hole circle and both the index plate and crank are turned 4 holes in a 12-hole circle.

These two movements will turn the worm through,

$$\frac{3}{15} + \frac{4}{12} + = \frac{8}{15}$$

Since 40 turns of the worm turn the work piece through one revolution, 8115 turns will move the work piece through,

$$\frac{8}{15} \times \frac{1}{40} = \frac{1}{75} \text{revolutions}$$

Thus, the work will be divided into 75 parts.

Let us take another example in which the index plate and the crank are rotated in the re-verse direction.

Suppose the crank is turned through 6 holes in the 18-hole plate and both the crank and crank pin arc turned in the reverse direction through 4 holes in the 16-hole plate. Then the worm will turn through,

$$\frac{6}{18} - \frac{4}{16} = \frac{1}{12}\,\text{revolutions}$$

Due to the above two movements, the work will move through,

$$= \frac{1}{12} \times \frac{1}{40} = \frac{1}{480}$$

Thus, the job can be divided into 480 pans. The following procedure is used for compound indexing:

- Factories the standard numbers 40 and the number of divisions.

- Select two circles on the index plate and factories their difference.

- Factories the number of holes of both the circles.

Place the factors as shown below:

$$= \frac{\text{Factors of division required} \times \text{factors of difference of hole circle}}{\text{Factors of 40} \times \text{factors of first circle} \times \text{factors of second circle}}$$

Suppose we get 1/x after simplification, If a and b are the number of holes on the two circles, the indexing movement is,

$$\frac{x}{a} - \frac{x}{b} \ \text{or} \ \frac{x}{b} - \frac{x}{a}$$

Activate after finding the values of a and b, check the algebraic sum of the two movements by the formula,

$$\frac{x}{a} + \frac{x}{b} = \frac{40}{N}$$

The dividing heads are of three types:

- Plain or simple dividing head.

- Universal dividing head.

- Optical dividing head.

(a) Plain or Simple Dividing Head

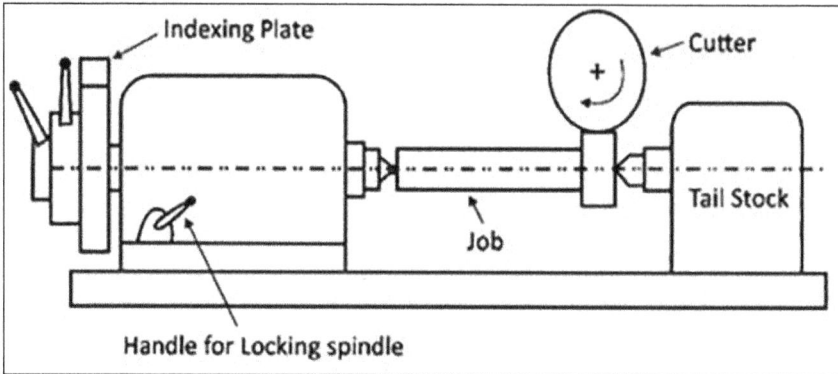

Plain Indexing Head for Direct Indexing.

The plain dividing head consists of a cylindrical spindle housed in a frame and a base bolted to the machine table. The indexing crank is connected to the tail end of the spindle directly and the crank and spindle rotate as one unit.

The index plate is mounted on the spindle and rotate with it. The spindle can be rotated through the desired angle and then clamped by inserting the clamping lever pin into any one of the slots of the index plate. The job is held between two centers, one on the dividing head spindle and the other on the tail stock as shown in figure above. In this dividing head, there is no worm and worm wheel.

(b) Universal Dividing Head

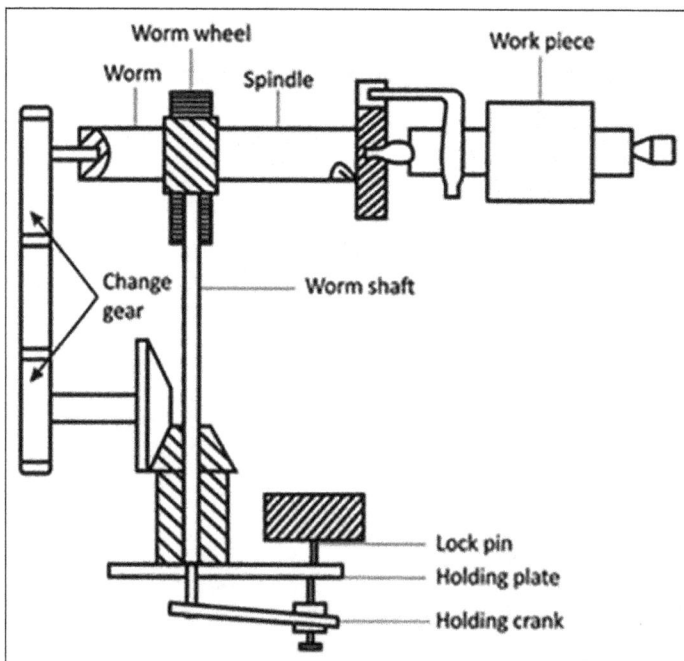

Working Mechanism of a Universal Dividing Head.

This dividing head is very useful device for the purpose of indexing work. The working mechanism of UDH is shown in Figure. The spindle carrying the worm wheel meshes with the worm, which carries a crank at its outer end. The worm wheel has 40 teeth and the worm is single threaded.

Thus 40 turns of crank will rotate the spindle for one complete revolution or one turn of the crank will cause the spindle to be rotated by 1/40 of a revolution. In order to turn the crank a fraction of a revolution of an index plate is used.

Index plate is a circular disc having a different number of equally spaced holes which is arranged in concentric circles. The index plate is screwed on a sleeve, which is loosely mounted on the worm shaft. Normally the index plate remains stationary by a lock pin. The index pin works inside the spring loaded plunger. This plunger can slide, radially along a desired hole circle on the index plate.

The dividing head spindle may be connected with the table feed screw through a train of gears to impart a continuous rotary motion to the work piece for helical milling.

(c) Optical Dividing Head

The optical dividing heads are used for precise angular indexing during machining and for checking the accuracy of various angular surfaces.

5.3.1 Differential and Angular Indexing Calculations

Differential Indexing

Differential indexing greatly resembles com-pound indexing. This process is also carried out in two stages. In the first operation, a crank is moved in a certain direction. In the second phase, movement is added or subtracted by moving the plate by means of a gear train.

For differential indexing, the dividing heads are supplied with standard sets of change gears. Brown and Sharpe supplied the dividing head with the following change gears. 24 (2 nos.), 28, 32, 40, 44, 48, 56, 64, 72, 86, 100 teeth.

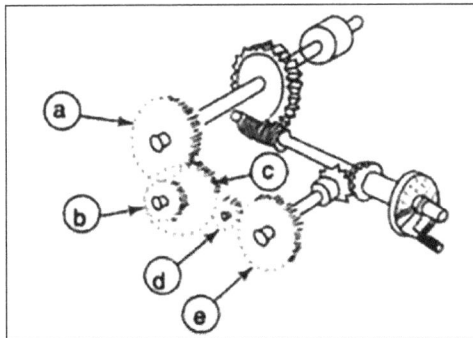

Differential Indexing. a b. c and e are inter-changeable gears.
d is idler gear. Index plate can move either in same or in opposite direction.

Depending upon the number of teeth to be cut, both simple and compound gear trains are used for differential indexing. In a simple gear train, motion is gained or lost depending on whether one or two idler wheels are used. In a compound gear train, the corresponding numbers are 0 and 1.

Angular Indexing

In a dividing head, the gear train is such that 40 revolutions of the crank rotate the workpiece through one revolution. In other words, 40 turns of the crank rotate the job through 360°. Thus, one turn of the crank rotates the job through 9°.

Now let us consider a 10-hole circle. Advancement of the crank through one hole will rotate the job through 9/10 degree or 54'. Similarly, advancement of the crank through one hole on 8, 9, 12, 15 and 18-hole circles will move the job through 9/8 (1°7' 30"), 9/9 (1°), 9/12 (0°45'), 9/15 (36') and 9/18 (30'), respectively.

Problem

1. Let us determine the differential indexing for 121 divisions.

Solution:

Selecting the suitable number of divisions, say 120,

$$\text{Simple indexing} = \frac{40}{120} = \frac{1}{3} = \frac{6}{18}$$

i.e. 6 holes on a 18-hole circle.

$$\text{Movement of crank for 121 divisions} = 121 \times \frac{1}{3} = 40\frac{1}{3}$$

i.e. 1/3 more than 40 revolutions.

Since the movement required is more, it is to be decreased by 1/3 revolution from the required 40 revolutions through plate movement.

$$\text{The gearing ratio} = \frac{1}{3} = \frac{1}{3} \times \frac{16}{6}$$

$$= \frac{16}{48} = \frac{\text{Driver}}{\text{Driven}}$$

Results

- Use a simple gear train.

- Driver with 16 teeth and driven with 48 teeth.

- Crank movement: 6 holes on a 18-hole circle.

5.3.2 Simple Problems on Simple and Compound Indexing

Simple Indexing

1. Let us describe the procedure of indexing 5 divisions on a work piece.

Solution:

Let, A = Number of turns through which the index crank is to be rotated.

N = Number of divisions required (= 5) Using the equation,

$$A = \frac{40}{N} = \frac{40}{5} = 8$$

Thus, for 1 division of the job, the crank should be rotated through 8 turns.

2. Let us do the indexing to cut 30 teeth on a spur gear blank which need us to divide the circumference of gear blank into 30 identical, parts.

Solution:

$$\text{Crank movement} = \frac{40}{N} = \frac{40}{30}$$

Here, N = 30,

$$= 1\frac{10}{30} = 1\frac{1}{3}$$

Let us multiply both numerator and denominator by 5.

$$= 1\frac{5}{15}$$

Denominator becomes '15' so we will select 15 hole circle of plate 1.

Action 1:

After each milling operation we will rotate indexing crank by one complete turn and 5 holes in 15 holes circle. This way we do milling for total 30 times.

In this case we can multiply numerator and denominator by '7 'in the place of '5'as described below.

$$\text{Indexing crank movement} = \frac{40}{N} \ (N = 30 \, \text{teeth})$$

$$= \frac{40}{30} = 1\frac{10}{30} = 1\frac{1}{3} \times \frac{7}{7} = 1\frac{7}{21}$$

Action 2:

We will select the hole circle of 21 holes. After each milling operation indexing crank will be rotated by 1 complete circle and 7 holes in 21 holes circle. This way milling operation will be done by total 30 times. Both the answers determined in the above problem are correct and substitute of each other.

Compound Indexing

3. Let us determine the compound indexing for 57 divisions.

Solution:

Required movement of the work piece $= \dfrac{40}{57}$.

Suppose we select two circles of 18 and 19 holes. Substituting the values in the expression.

$$\frac{\text{Divisions required} \times \text{difference of hole cicle}}{40 \times \text{no. of holes of first circle} \times \text{no. of holes of second circle}}$$

$$= \frac{57 \times 1}{40 \times 18 \times 19} = \frac{1}{240}$$

Since the numerator is unity, the circles selected are correct. The required indexing movement is given by,

$$\frac{240}{18} - \frac{240}{19} \text{ or } \frac{240}{19} - \frac{240}{18}$$

or

$$13\frac{6}{18} - 12\frac{12}{19} \text{ or } 12\frac{12}{19} - 13\frac{6}{18}$$

Taking 13 as common, the above expression becomes,

$$\frac{6}{18} + \frac{7}{19} \text{ or } -\frac{7}{19} - \frac{6}{18}$$

Similar signs means that the movement will be in the same direction. The above expression shows that the crank will move 6 holes in a 18-plate circle and the crank and the index plate will move 7 holes in a 19-hole circle to get 57 divisions.

$$\text{Check} \frac{6}{18} + \frac{7}{19} = \frac{114 + 126}{342} = \frac{240}{342}$$

$$= \frac{40}{57} \left(\frac{40}{N} \right)$$

4. Let us make 69 divisions of work piece circumference by indexing method. (Using compound indexing).

Solution:

Follow the steps given below:

1. Factor the divisions to be made (69 = 3x23) N = 69.

2. Select two hole circles at random (These are 27 and 33 in this case, both of the hole circles should be from same plate).

3. Subtract smaller number of holes from larger number and factor it as (33 − 27 = 6 = 2x3).

4. Factor the number of turns of the crank required for one revolution of the spindle (40). Also factorize the selected hole circles.

5. Place the factors of N and difference above the horizontal line and factors of 40 and select both the hole circles below the horizontal line as given below. Cancel the common values.

6. If all the factors above the line are cancelled by those which are below the line, then the selected hole circles can be used for indexing otherwise select another two hole circles. In this case there is need to select another hole circles. Let us select 23 and 33 this time and repeat the step 5 as indicated below,

$$69 = 23 \times 3$$
$$\underline{6 = 2 \times 3}$$
$$40 = 2 \times 2 \times 2 \times 5$$
$$27 = 3 \times 3 \times 3$$
$$33 = 3 \times 11$$

Encircled numbers below the line are the left out numbers after canceling the common factors. All the factors above the horizontal line are cancelled so selected hole circles with 22 and 33 holes can used for indexing.

$$69 = 23 \times 3$$
$$\underline{10 = 2 \times 5}$$
$$40 = 2 \times 2 \times 2 \times 5$$
$$22 = 23 \times 1$$
$$33 = 11 \times 3$$

7. Following formula is used for indexing:

$$\frac{40}{69} = \frac{n_1}{N_1} + \frac{n_2}{N_2}$$

In this formula $N_1 = 23$ and $N_2 = 33$ (N_1 is always given smaller value out of two.)

8. Multiply all the remaining factors below the line as 2 x 2 x 11 = 44. The formula above will turn to,

$$\frac{40}{69} = \frac{44}{23} - \frac{44}{33}$$

We will neglect the + ve sign.

$$= 1\frac{21}{23} - 1\frac{11}{33}$$

The −ve sign indicates backward movement.

Action:

For indexing of 69 divisions, the indexing crank should be moved by 21 holes circle in forward direction and then crank along with the plate are moved by 11 holes in 33 hole circle in reversed (backward) direction.

Grinding Machines

6.1 Types of Abrasives

Basically there are two types of abrasives used in the manufacturing of a grinding wheel. These are:

- Natural Abrasive: Such as sand stone, emery, corundum and Diamonds.

- Synthetic Abrasives: These are artificially manufactured abrasives which are mostly used in grinding wheels. Synthetic abrasives are:

  - Silicon carbide.

  - Aluminium oxide.

  - Boron carbide.

  - Boron Nitride.

  - The Diamonds.

Natural Abrasives

Natural abrasives such as corundum and emery have long been used for grinding operations. These are seldom used in production and precision work. These are mostly used as an abrasive coating on cloth and paper and are popularly known as emery cloth and emery paper. These are also employed in buffing operations. Sandstone is used for cutting and polishing of floors.

Synthetic Abrasives

The Diamonds

Only the natural diamond stones which cannot be converted into invaluable gems known as Bort are used for machining operations. The powdered form of such natural stone (Bort) which is the hardest known substance is used in the manufacturing of diamond grinding wheels and other diamond cutting tools.

The diamond abrasive grinding wheels are employed for providing tool geometry in cemented carbide, tungsten carbide, ceramic and diamond cutting tools and giving shape to other diamond products and diamond glass cutter.

For manufacturing and production work the grinding wheels used are made of synthetic abrasives.

Silicon Carbide

Silicon carbide is produced by fusing silica sand, coke, saw dust and salt. Powdered and graded abrasive particles are used in manufacturing grinding wheels. Silicon carbide grinding wheels give excellent results when used for softer materials, such as iron, brass, bronze, copper, aluminium. Silicon carbide being harder and brittle is also used for grinding cemented carbide, glass and ceramics.

Aluminium Oxide

Aluminium oxide is made by fusing bauxite and is slightly softer than silicon carbide but it is very tough and is mostly used for grinding hard materials having high tensile strength.

Boron Carbide and Boron Nitride

Boron carbide and boron nitride abrasive grains are only next to diamond in hardness. These are bonded under high temperature and high pressure, as a result, these are very dense in construction. Such grinding wheels are usually employed for grinding tool and die steel, alloy steel, stainless steel, forged steel, mild steel, cast iron and super alloys.

6.1.1 Grain Size, Grade and Structure

The size of the abrasive grains is indicated by a grain or grit size. Grinding is a true cutting operation and the grain size also indicates the size of the cutting teeth. Grain size is the mesh size, i.e., the number of meshes per linear inch (25.4 mm) of a sieve through which the grains pass when graded after crushing.

Table: Common grain size and type.

SI. No	Grain Type	Grain size						
1	Coarse	10	12	14	16	20	-	-
2	Medium	30	36	46	54	60	-	-
3	Fine	80	100	120	150	180	-	-
4	Very fine	220	240	280	320	400	500	600

The common grain sizes and types used for the manufacture of grinding wheels arc given in the table. The grain size used in a grinding wheel depends on the amount of metal to be removed, the desired finish and hardness of work material.

Example: Course wheels are used for fast material removal. Fine wheels are used for soft ductile materials. These should also be used for hard and brittle work-material.

Grade

The grade of a grinding wheel refers to the hardness with which the wheel holds the abrasive grains in place. It does not refer to the hardness of the abrasive grains. The term 'soft' or 'hard' refers to the resistance a bond offers to disruption of the abrasives.

A wheel from which the abrasive grains can easily be dislodged is called soft whereas the one, which holds the grains more securely, is called hard. The grade of the bond can be classified in three categories.

Table: Suitable grade depending on the grinding condition.

Low (Soft)	Grade	High (Hard)
Hard and/or fragile	Workpiece.	Soft and/or tough.
Wide	Contacting area.	Narrow.
High	Rotation speed of grinding wheel.	Low.
Low	Rotation speed of workpiece.	High.

Structure

The proportion of the volume of grain within the given volume of the entire grinding wheel is called grain ratio.

The grain ratio is distributed into 15 grades, namely from 0 to 14, which being the structure number of the grinding wheel. The larger the number, the lower the grain ratio i.e. larger space between grains. Space between grains has great influence on the grinding efficiency and incidence of heating.

Table: Grain ratio and structure number.

Structure number	0	1	2	3	4	5	6	7	8	9	10	11	12	13	14
Grain size	62	60	58	56	54	52	50	48	46	44	42	40	38	36	34

6.1.2 Bonding Process

Bonds and Bonding Processes: A bond is an adhesive substance that is employed to hold abrasive grains together in the form of sharpening stones or grinding wheels.

Bonding materials and processes are:

- Vitrified bond used for making verified grinding wheels.

- Silicate bond for making silicate wheels.

- Shellac bond for making elastic wheels.

- Resinoid bond used for making resinoid wheels.

- Rubber bond used for making vulcanized wheels.

- Oxychloride bond for making oxychloride wheels.

1. Vitrified Bonding Process:

Vitrified wheels are made by bonding clay melted to a glass like consistency with abrasive grains. The clay and abrasive grains are thoroughly mixed together with sufficient water to make the mixture uniform. The fluid mixture is then poured into molds and allowed to dry. When it has dried to a point where it can be handled, the material is cut trimmed to more perfect size and shape.

It is then heated or burned in a kiln in much the same manner as brick or tile is burnt. When the burning proceeds, the clay vitrifies, that is, it fuse and forms a porcelain or glass like substance that surrounds and connects the abrasive grains. The high temperature employed in this process tends to anneal the abrasives to some extent.

Vitrified bond gives a wheel good strength as well as porosity to allow high stock removal with cool cutting. It is affected by heat, cold water or acids.

Disadvantages of vitrified bonded wheels are their sensitivity to impact and their low bending strength. About 75% of the wheels now manufactured are made with this bond. A vitrified bonded wheel is denoted by the letter 'V'.

2. Silicate Bonding Process:

Silicate wheel are made by mixing abrasive grains with silicate of soda or water glass. The mixture is packed into molds and allowed to dry. The molded shapes are then backed in a furnace of a temperature of 260°C for several days.

The silicate bond releases the abrasive grains more readily than the vitrified bond, the abrasive grains are not annealed as in the vitrified process and silicate wheels are waterproof ,These characteristics make silicate wheels valuable for grinding edged tools and other operations where heat must be held to a minimum with or without the aid of a coolant. A silicate bonded wheel is denoted by the letter 'S'.

3. Shellac Bonding Process:

Shellac bonded wheels are also known as elastic bonded wheels. In this process, the abrasive and shellac are mixed in heated containers and then rolled or pressed in heated molds. Later the shapes are backed a few hours at a temperature of approximately 150 °C.

The elasticity of this bond is greater than in other types and it has considerable strength. It is not intended for heavy duty. Shellac bond is cool cutting on hardened steel and thin sections and is used for finishing chilled iron, cast iron and steel rolls, hardened steel cams and aluminum pistons and in very thin sections, for abrasive cutting of machines. A shellac bonded wheel is denoted by the letter 'E'.

4. Resinoid Bonding Process:

Resinoid wheels are produced by mixing abrasive grains with synthetic resins and other compounds. The mixture is placed in molds and heated at about 200°C. At this temperature, the resin sets to hold the abrasive grains in wheel form.

Wheels bonded with synthetic resin, such as Bakelite and Redmanol, are used for purposes which require a strong, free high speed wheel. They can remove stock Very rapidly. They are useful for precision grinding cams and rolls requiring high finish. A resinoid bonded wheel is denoted by the letter 'B'.

5. Rubber Bonding Process:

Rubber bonded wheels are prepared by mixing abrasive grains with pure rubber and sulfur. The mixture rolled into sheets and wheels are punched out of the sheets on a punch press. Following that, the wheels are vulcanized.

6. Oxychloride Bonding Process:

This process consists of mixing abrasive grains with oxide and chloride of magnesium:

- The mixing of bond and abrasive is performed in the same way as for vitrified bond wheel.

- These wheels are used in making wheels and wheel segments for disc grinding operation.

6.2 Grinding Wheel Types

A grinding wheel is a multi-tooth cutter made up of many hard particles known as abrasives which have been crushed to leave sharped edges for machining.

Every grinding wheel has two constituents:

- Abrasive used for cutting.

- Bond that holds abrasive grains.

Basic Functions of a grinding wheel:

- Removal of stock.

- Generation of cylindrical, flat and curved surfaces.

- Production of highly finished surfaces.

- Cutting off operations.

- Production of sharp edges and points.

Construction of Grinding Wheel

Grinding wheel consists of:

- Abrasives

- Bond

- Grit/grain size

- Grade

- Structure of wheels

There is different type of grinding wheel:

- Straight grinding wheels: Straight wheel are the most common mode of wheel that is found on pedestal or bench grinders. This is the one widely used for center less & cylindrical surface grinding operations. As it is used only on the periphery, it forms a little concave surface on the piece. This is used to gain on several tools like chisels. The size of these wheels differs to a great extent, width & diameter of its face obviously depends on the category of its work, machines grinding power.

- Cylinder or wheel ring: A cylinder wheel has no center mounting support but has a long & wide surface. Their width is up to 12" and is used purely in horizontal or vertical spindle grinders. This is used to produce flat surface, here we do grinding with the ending face of the wheel.

- Tapered Grinding wheels: Tapered Grinding wheel is a straight wheel that tapers externally towards the midpoint of the wheel. As this pact is stronger than straight wheels, it accepts advanced lateral loads. Straight wheel with tapered face is chiefly used for gear teeth, grinding thread, etc.

- Straight cup: This Straight cup wheels forms an option for cup wheels in cutter and tool grinders, having an extra radial surface of grinding is favorable.

- Dish cup: In fact this is used primarily in jig grinding and cutter grinding. It is a very thin cup-style grinding wheel which permits grinding in crevices and slot.

- Saucer Grinding Wheels: Saucer Grinding Wheel is an exceptional grinding profile used for grinding twist drills and milling cutters. This finds wide usage in non-machining areas, as this saw filers are used by saucer wheels to maintain saw blades.

- Diamond Grinding Wheels: In diamond wheels industrial diamonds remain bonded to the edge. This is used to grind hard materials like concrete, gemstones and carbide tips. A slitting saw is designed for slicing gemstones like hard materials.

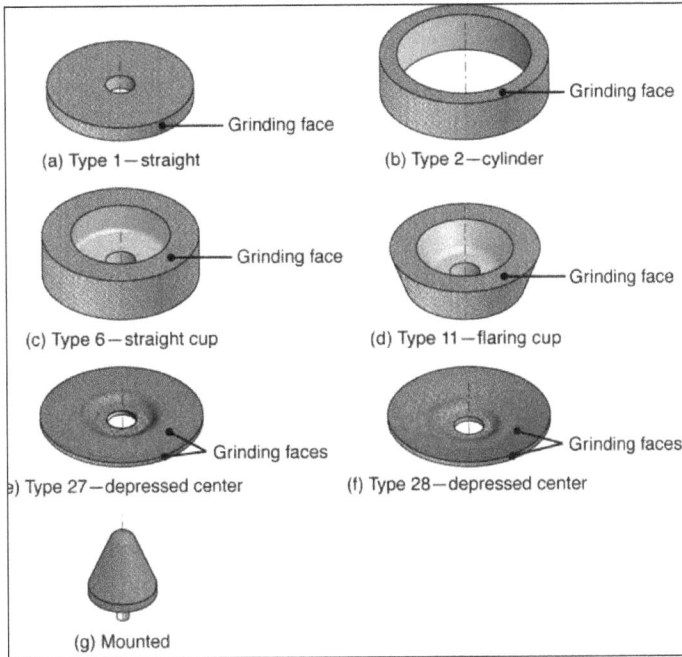

(a) Type 1 — straight
Grinding face

(b) Type 2 — cylinder
Grinding face

(c) Type 6 — straight cup
Grinding face

(d) Type 11 — flaring cup
Grinding face

(e) Type 27 — depressed center
Grinding faces

(f) Type 28 — depressed center
Grinding faces

(g) Mounted

Types of grinding wheels.

Specification of Grinding Wheel

1. Geometrical Specification

This is decided by the type of grinding machine and the grinding operation to be performed in the work piece. This specification mainly includes wheel diameter, width and depth of rim and the bore diameter.

The wheel diameter, for example can be as high as 400mm in high efficiency grinding or as small as less than 1mm in internal grinding. Similarly, width of the wheel may be less than an mm in dicing and slicing applications. Standard wheel configurations for conventional are shown in figure below.

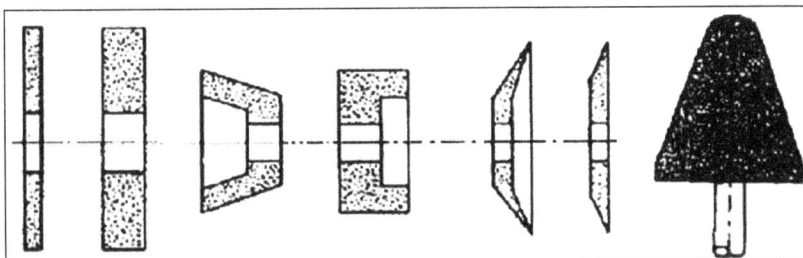

Standard wheel configuration for conventional grinding wheels.

2. Compositional Specifications

Specification of a grinding wheel ordinarily means compositional specification. Conventional abrasive grinding wheels are specified encompassing the following parameters.

- The type of grit material.

- The grit size.

- The bond strength of the wheel, commonly known as wheel hardness.

- The structures of the wheel denoting the porosity i.e. the amount of inter grit spacing.

- The type of bond material.

- Other than these parameters, the wheel manufacturer may add their own identification code.

- Prefixing or suffixing (or both) the standard code.

6.2.1 Classification, Constructional Features of Grinding Machines

Classification of Grinding Machine

There are various types of grinding machines which have been designed and are used. Some of them are used for roughing work, for precision work and for special purposes. However, the most commonly used types can be broadly classified as follows:

- Rough or non-precision grinders.

- Precision grinders.

1. Rough or Non-Precision Grinders:

The main purpose of roughing grinders is to remove more material than that can be removed by the other types of grinders. This class of grinders include following grinding machines:

- Bench, pedestal or floor grinders.

- Swing frame grinders.

- Portable grinders.

- Belt grinders.

2. Precision Grinders:

Precision grinders remove a small amount of material and finish the work piece to a very accurate dimension.

As per the type of surface generated, precision grinders are classified as follows:

Cylindrical grinders:

- Plain cylindrical grinders.
- Universal cylindrical grinders.
- Center-less grinders.

Surface Grinders:

- Reciprocating table:
  - Horizontal spindle.
  - Vertical spindle.
- Rotary table:
  - Horizontal spindle.
  - Vertical spindle.

Internal Grinders:

- Plain internal grinders.
- Universal internal grinders.
- Planetary internal grinders.
- Center less internal grinders.

Methods of Cylindrical Grinding:

- Center-type (plain).
- Center-type (universal).
- Center-less.

Centerless Type Grinders

Centerless grinding is a method of grinding exterior cylindrical tapered and formed surfaces on work pieces that are not held and rotated on carters. The principal elements of an external centerless grinder shown in figure are the grinding wheel, regulating wheel and the work rest. Both wheels are rotated in the same direction. The work rest is located between the wheels.

The work is placed upon the work rest and the latter, together with the regulating wheel, is fed forward, forcing the work against the grinding wheel.

A. Grinding wheel
B. Grinding face
C. Regulating wheel
D. Work piece
E. Work rest blade

θ = Angle of tilt of
regulating wheel

Movements
1. Grinding wheel   2. Work
3. Regulating wheel   4. Infeed
5. Traverse

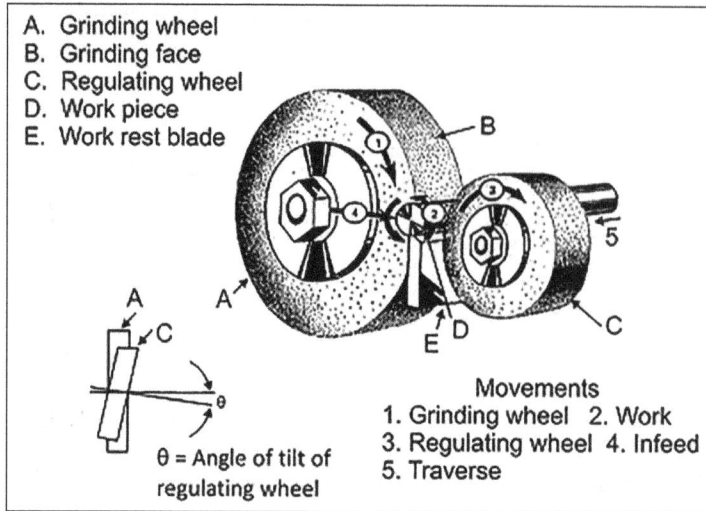

Centerless type grinders.

Centerless grinding may be done in one of the three ways:

- Through feed.

- In Feed.

- End feed.

1. Through feed Grinding: The work is passed completely through the space between the grinding wheel and regulating wheel, usually with guides at both ends. This method is used when there are no shoulders or other form to interfere with the passage of the work. It is useful for grinding long, slender bar.

2. In Feed Grinding: Which is similar to plunge grinding or form grinding, the regulating wheel is drawn back and the work piece may be placed on the work rest blade. Then it is moved in to feed the work against the grinding wheel. This method is used to grind shoulder, formed surfaces.

3. In End Feed Grinding: Used to produce tapes, either the grinding wheel or regulating wheel or both are formed to a taper. The work is fed lengthwise between the wheels and is ground as it advances until it reaches the end stop.

Centerless Grinding

The external center less grinding principle is also applied to internal grinding. In internal center less grinding, the work is supported by three rolls.

One is the regulating wheel, the second one is a supporting roll and the last one is pressure roll to hold the work piece against the support and regulating rolls. The regularity roll is a rubber bonded wheel. This roll makes the work piece to rotate.

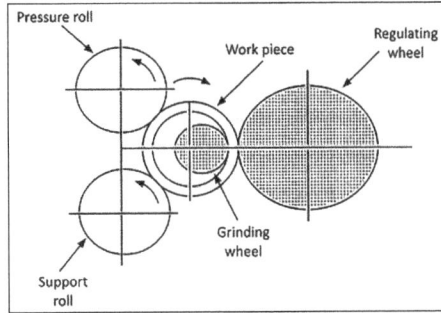

Centerless grinding.

The grinding wheel contact inside diameter of the work piece directly and reciprocates about its axis for giving the feed. The depth of cut is given by moving the grinding wheel in a crosswise direction. The pressure roll is mounted to swing aside to permit loading and unloading.

Centerless grinding is a method of grinding exterior cylindrical, tapered and formed surfaces on work pieces that are not held and rotated on centers. The principal elements of an external center less grinder shown in figure are the grinding wheel, regulating or back up wheel and the work rest. Both wheels are rotated in the same direction.

The work rest is located between the wheels. The work is placed upon the work rest and the latter, together with the regulating wheel, is fed forward, forcing the work against the grinding wheel.

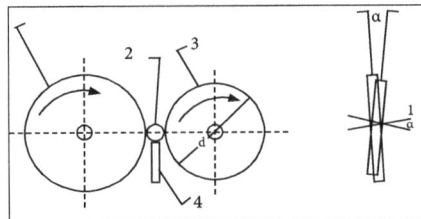

External centerless grinding:

- Grinding wheel.

- Work.

- Regulating wheel.

- Work-rest.

The axial movement of the work past the grinding wheel is obtained by tilting the regulating wheel at a slight angle from horizontal. An angular adjustment of 0 to 8 or 10 degrees is provided in the machine for this purpose. The actual feed (s) can calculate by the formula:

$$s = \pi d n \sin \alpha$$

Where,

s = Feed in mm per minute.

n = Revolution per minute.

d = Diameter of regulating wheel in mm.

α = Angle of inclination of wheel.

The advantages of centerless grinding are:

- As a true floating condition exists during the grinding process, less metal needs to be removed.

- The work piece being supported throughout its entire length as grinding takes place, there is no tendency for chatter or deflection of the work and small, fragile or slender work pieces can be ground easily.

- The process is continuous and adapted for production work.

- No center holes, no chucking or mounting of the work on mandrels or other holding devices are required.

- The size of the work is easily controlled.

- A low order of skill is needed in the operation of the machine.

Plain Center Type Cylindrical Grinding

A plain center type cylindrical grinding machine is shown in figure. These grinding machines are used for grinding mainly cylindrical parts.

They are also used for grinding such as tapers, fillets, contoured cylinders etc. The grinding machine consists of various parts.

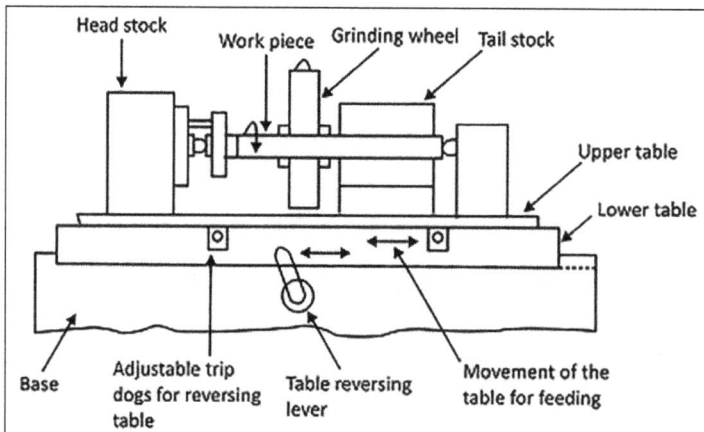

Plain center type cylindrical grinding machine.

1. Base: The base is the main casting that rests on the floor and supports the parts mounted on it. On the top of the base, horizontal guide ways are set on which the table slides to give traverse motion to the work piece. The table drive mechanism is incorporated in the base itself.

2. Table: There are two tables such as upper table and lower table. The lower table slides on the guide ways of the bed and provides traverse feed or longitudinal feed of the work past the grinding wheel. It can be moved by hand or power within the limits.

3. Head Stock: The head stock supports the work piece by means of a dead center. The work piece is driven by head stock dog and driving pin.

4. Tail Stock: The tail stock can be adjusted and clamped to accommodate different length of work pieces. The work piece is held in between the center of head stock and tail stock.

5. Wheel Hard: It carries a grinding wheel and rotated by a motor housed in the hand stock. The wheel hand is placed over the bed at its backside.

Traverse Grinding

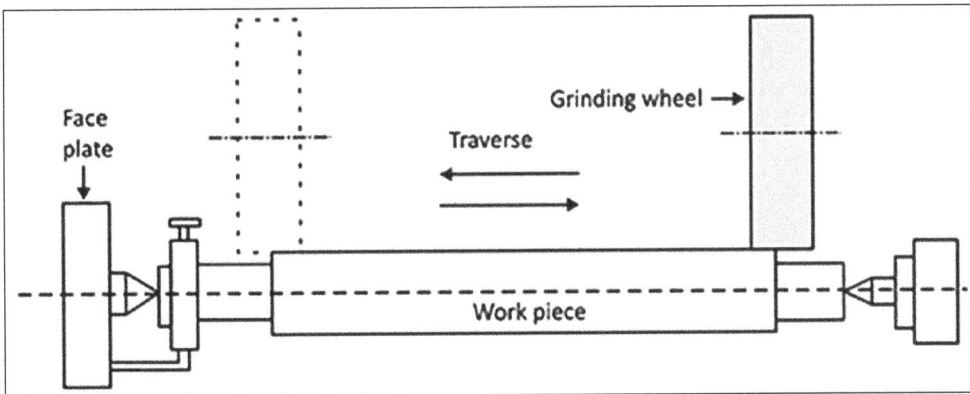

Traverse Grinding.

This method is used when the job length is more than the width of the grinding wheel. The job is held between two centers. The grinding wheel is made to rotate in a fixed position. The rotating work is made to traverse. The rotating work moves longitudinally in both directions. This is longitudinal feed.

Plunge Grinding

This method is used when the length of the work piece is lesser than the width of the grinding wheel. Here, the work piece need not be fed longitudinally. The grinding is done by giving only the cross feed to the grinding wheel. This is known as plunge grinding. Here, no longitudinal movement is given to the work piece. Therefore, the table is locked. The wheel is fed till the required diameter is obtained.

The wheel is specially shaped. Plunge grinding is used for grinding shoulders, stepping and various contours on the work piece.

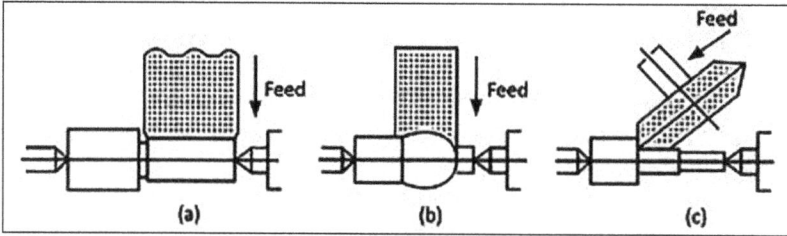

Plunge Cut Grinding.

6.2.2 Surface Grinding

Surface-grinding machines are generally used for generating flat surfaces. By far, these are the largest amount of grinding work done in most of the machine shops. These machines are similar to milling machines in construction as well as motion.

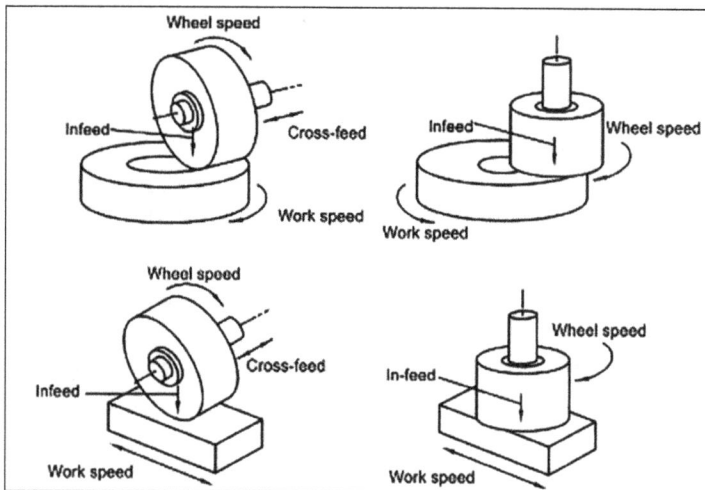

Surface grinding.

There are basically four types of machines depending upon the spindle direction and the table motion as shown in the figure. They are:

- Horizontal spindle and rotating table.

- Vertical spindle and rotating table.

- Horizontal spindle and reciprocating table.

- Vertical spindle and reciprocating table.

Horizontal Spindle and Rotating Table

In this machine, the grinding wheel cuts on its periphery, while the spindle traverses horizontally from the edge to the centre of the table. Feed is accomplished by moving the work mounted on the table, up into the wheel, with the table moving in a rotary fashion. Since the table and work rotate in a circle beneath the grinding wheel, the surface pattern is a series of intersecting arcs. This machine is used for round, flat parts because the wheel is in contact with the work at all times.

Vertical Spindle and Rotating Table

Vertical spindle machines are generally of bigger capacity. The complete machining surface is covered by the grinding-wheel face. They are suitable for production grinding of large flat surfaces. In this machine, both the work and the wheel rotate and feed into each other.

By taking deep cuts, this machine removes large amounts of material in a single pass. The side or the face of the wheel does the grinding. The wheel can be either complete solid or split into segments to save wheel material and in the process also provide cooler grinding action.

In the case of small parts, the surface patterns created are a series of intersecting arcs if they are off-centre around the table. It is a versatile machine and can be used to grind production pans and very large parts, as well as for grinding large batches of small parts as shown in the figure above.

Horizontal Spindle and Reciprocating Table

The table in the case of reciprocating machines is generally moved by the hydraulic power. The wheel head is given a cross-feed motion at the end of each table motion. In this machine, the wheel should over travel the work piece at both the ends to prevent the grinding wheel, removing the metal at the same work spot during the table reversal. This is the most common grinding machine found in tool rooms.

The tables for this type of machines are rectangular and usually 150 mm wide by either 300 mm or 450 mm long. The high-production types have tables as big as 2 in by 5 m. The grinding wheels cut on their peripheries and vary in sizes from 175 mm in diameter and 12.5 mm in width to 500 mm in diameter and 200 mm in width.

This type of surface grinder is the most commonly used because of its high accuracy and the fine surface finishes that it imparts. The grinding wheel traverses in a straight pattern that results in superior finish and high precision.

Vertical Spindle and Reciprocating Table

The grinding wheel in this machine is cylindrical and cuts on its side rather than on its periphery. The work is fed by the reciprocating motion of the table. Generally, the diameter of the wheel is wider than the work piece and as a result no traverse feed is required. These are generally high-production machine tools removing large amounts (as much as 10 mm) in a single pass.

Internal Grinding:

- The work piece is chucked and rotated about its axis. The work head is mounted at the left side of the machine. The wheel head is mounted at the right end of the machine. The grinding wheel is rotated.

- At the same time, it reciprocates back and forth through the length of the hole as shown in figure.

- These machines are used for grinding work piece which can be easily held in a chuck.

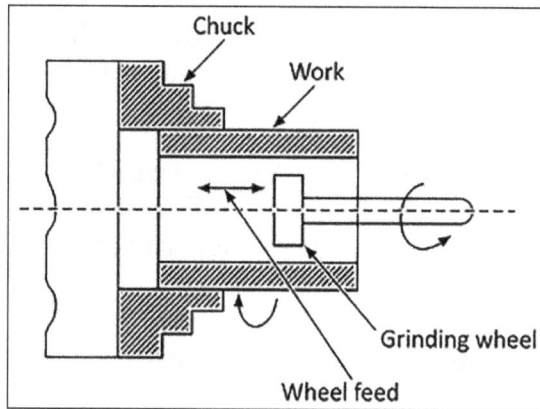

Chucking type internal grinder.

Advantages:

- Dimensional accuracy.

- Good surface finish.

- Good form and locational accuracy.

- Applicable to both hardened and unhardened material.

Applications:

- Surface finishing.

- Slitting and parting.

- Descaling, deburring.

- Stock removal (abrasive milling).

- Finishing of flat as well as cylindrical surface.

- Grinding of tools and cutters and resharpening of the same.

Concepts of Surface Integrity

Surface integrity is the surface condition of a work piece after being modified by a manufacturing process. The surface integrity of a work piece or item changes the material's properties. The consequences of changes to surface integrity are a mechanical engineering design problem, but the preservation of those properties are a manufacturing consideration.

6.3   Selection of Grinding Wheel

Factors to be considered to select a grinding wheel and recommended parameters:

- The proper grain size, bond, strength, grade, shape and size of the wheel should be selected to meet the specific requirements.

- The proper selection of a grinding wheel is important to ensure rapid work, good surface finish and increased wheel life. To get optimum results, the various elements that influence the process need consideration.

The factors that influence the selection of a grinding wheel can be classified as:

- Constant factors

- Variable factors

Constant factors depend upon the material of the work piece, the amount of mate-rial to be removed, the area of contact and the finish and accuracy required.

Variable factors depend upon the speed of the grinding wheel, the speed of the work piece and the skill of the operator. Various factors that need consideration for selection of a grinding wheel are abrasives, grain size and shape, type of bond, bond strength and hardness. A brief description of these elements are as follows:

Abrasive

The selection of an abrasive depends upon the material to be ground. Silicon carbide (SiC) and aluminium oxide (Al_2O_3) are abrasives commonly used for grinding wheels. Silicon carbide is used for hard materials while aluminium oxide is used for soft materials.

Wheel Speed

The speed of a grinding wheel is influenced by the grade and the bond. The higher the speed of a grinding wheel, the softer it is. However, the speed of grinding wheels cannot be increased beyond permissible limits.

Work Speed

The speed at which the work piece traverses across the wheel face is known as the work speed. The higher the speed of work, the greater is the wear and tear of the wheel. If the work speed is low, the wheel wear is also low. However, low speed results in local over-heating, produces deformation and lowers the hardness of work pieces by producing tempering treatment.

Most grinding machines are provided with variable speed mechanisms. As the diameter of the wheel decreases, the speed needs to be increased accordingly to provide optimum working conditions.

6.3.1 Grinding Process Parameters

The figure (1a) indicates progressive decrease in grinding force with increase of grinding velocity. The opposite trend is observed when work piece traverse speed on wheel depth of cut is increased.

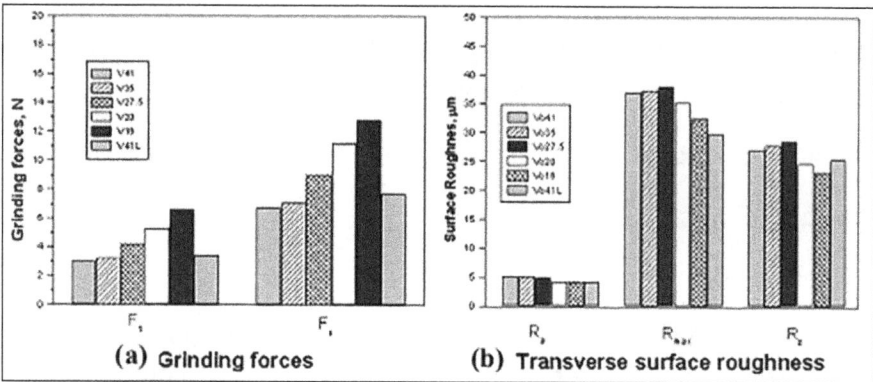

(a) Grinding forces　　　(b) Transverse surface roughness

(1) Effect of grinding velocity (m/s).

This is indicated in the figure (2a) and figure (3a). The variation of uncut layer thickness with grinding parameters causes the variation in force per grit as well as in total grinding force.

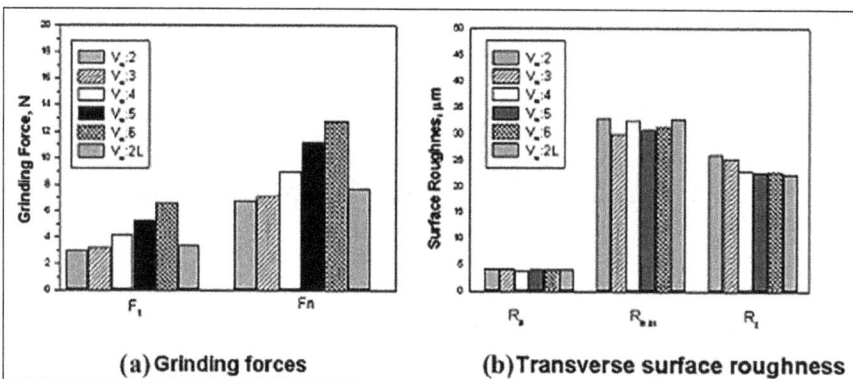

(a) Grinding forces　　　(b) Transverse surface roughness

(2) Effect of table feed (m/min).

Surface roughness of the work piece in the transverse direction is a subject of major concern. Surface roughness in longitudinal direction is mostly found to be significantly low. The transverse surface roughness of a work piece depends mainly on the grit geometry, overlap cuts made by the grits and lateral plastic flow of the work material.

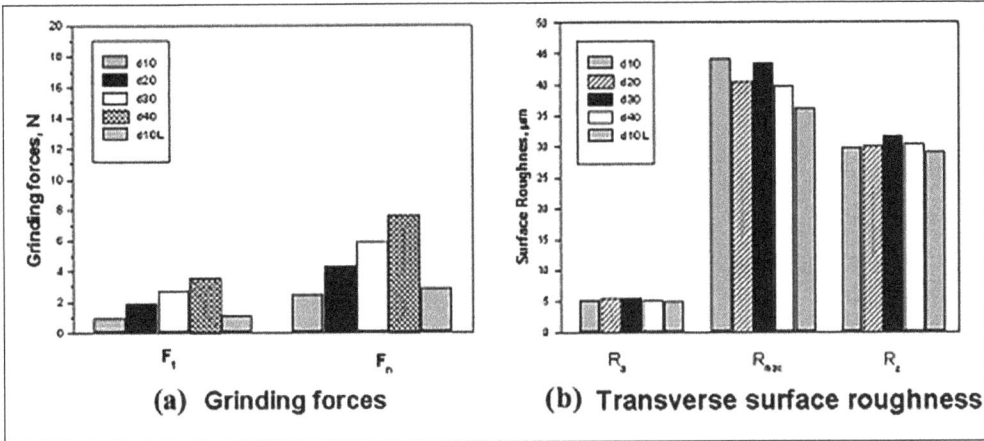

(3) Effect of depth of cut (μm).

Grinding parameters like grinding velocity, traverse speed or wheel depth of cut affects the grinding force which in turn can cause fracture, rounding or flattening on few overlying grits thus, bringing more number of underlying grits into action. This change in topographical feature of single layer wheel, in various levels, affects the surface roughness of the work piece as illustrated in the figure (1b), (2b) and (3b).

Grinding force increases with decrease in grinding velocity while the same increases with increase in table speed and depth of cut. Accordingly a trend is observed on decrease of surface roughness with decrease in grinding velocity and increase of both traverse speed and wheel depth of cut.

Grinding Points

Grinding Points are capable of reaching places inaccessible to larger types of grinding wheels. It can be used safely on hi-speed portable grinding machines.

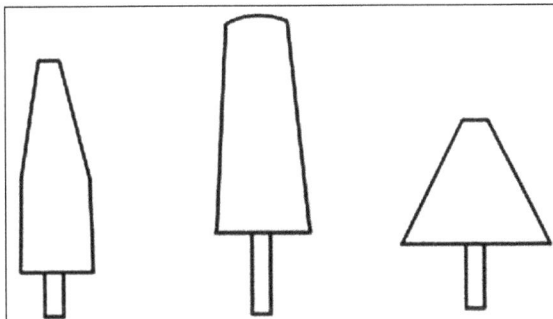

Grinding points.

6.3.2 Dressing and Truing of Grinding Wheels

Dressing

If the grinding wheels are loaded or gone out of shape, they can be corrected by dressing or truing of the wheels. Dressing is the process of breaking away the glazed surface so that sharp particles are again presented to the work. The common types of wheel dressers known as "Star" -dressers or diamond tool dressers are used for this purpose.

A star dresser consists of a number of hardened steel wheels on its periphery. The dresser is held against the face of the revolving wheel and moved across the face to dress the wheel surface. This type of dresser is used particularly for coarse and rough grinding wheels.

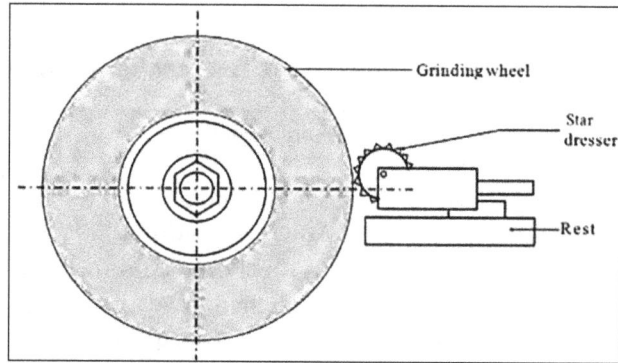

Dressing of a grinding wheel (star wheel method).

For precision and high finish grinding, small industrial diamonds known as 'bort' are used. The diamonds are mounted in a holder. The diamond should be kept pointed down at an angle of 15° and a good amount of coolant is applied while dressing. Very light cuts only may be taken with diamond tools.

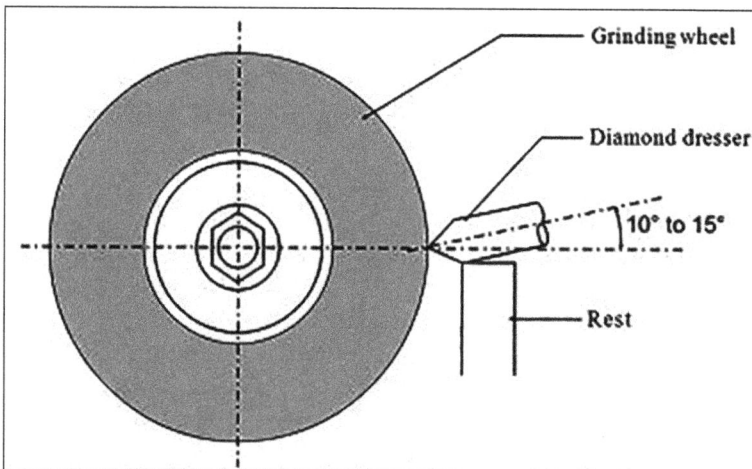

Dressing of a grinding wheel (Diamond dresser method).

Truing

The grinding wheel becomes worn from its original shape because of breaking away of the abrasive and bond. Sometimes the shape of the wheel is required to be changed for form grinding. For these purposes the shape of the wheel is corrected by means of diamond tool dressers. This is done to make the wheel true and concentric with the bore or to change the face contour of the wheel. This is known as truing of grinding wheels.

Diamond tool dressers are set on the wheels at 15° and moved across with a feed rate of less than 0.02mm. A good amount of coolant is applied during truing.

Broaching Process

7.1　Principle of Broaching

Broaching is a process of machining a surface with a special multi point cutting tool called broach which has successively higher cutting edges in a fixed path.

Broaching is a machining process in which metal removal takes place with the help of a number of successive teeth incorporated on the broach. Cutting takes place by a transverse cutting action, by pushing or pulling the broach through the hole or surface. Broaching is an efficient and rapid process, because roughing and finishing operations can be done in a single pass.

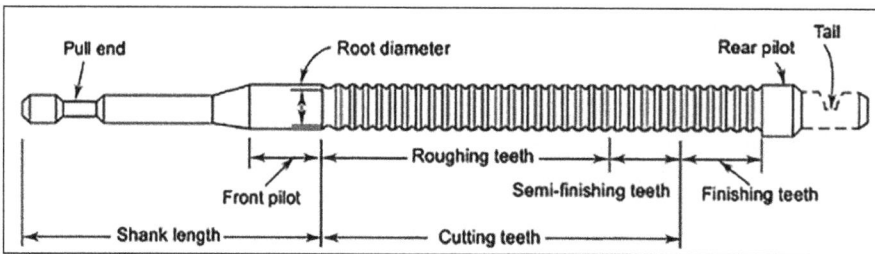

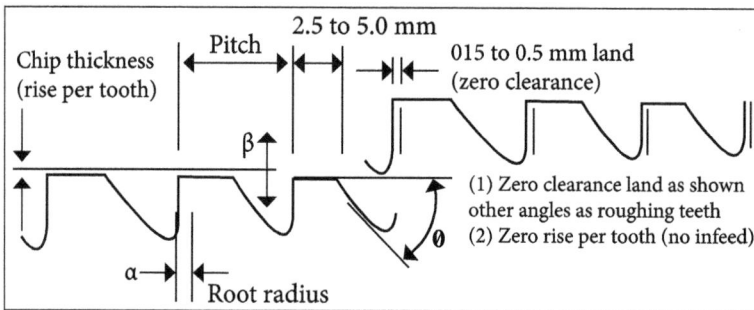

Broaching machine.

- Pull end for engaging the broach in the machine.

- Neck of shorter diameter and length, where the broach is allowed to fail, if at all, under overloading.

- Front pilot for initial locating the broach in the hole.

- Roughing and finishing teeth for metal removal.

- Finishing and burnishing teeth for fine finishing.

- Rear pilot and follower rest or retriever.

Principle of Operation of Broaching for Internal and External Machining

The work piece is set on the table by means of fixture. The broach is moved over the surface in a fixed path. The broach has multiple cutting teeth along its length. The height of tooth is gradually increasing. Each tooth removes a very small amount of metal.

The total metal removing is done progressively by each teeth. The machining is done in one stroke of broach. The broach may be pulled or pushed over the surface of work piece.

Detail of an internal or hole broach.

A typical broach is shown in figure. It is used to machine an internal hole. The broach is gripped by puller at the shank end. The front rake angle refers a rake angle of a single point cutting tool and back of the angle (relief angle) is provided to prevent rubbing of the tool with the work piece.

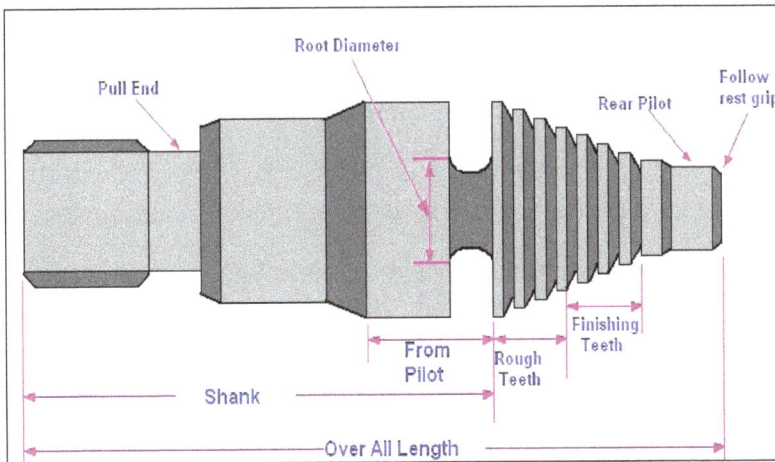

Internal or hole broach.

High Speed Steel

Material is widely used to make the broach. It is also raised carbide of disposable inserts or sometime used for cutting edges then machining cost iron parts, which requires close tolerance. Carbide tools are also used to an advantage on steel cutting. A broach may be either assembled or built up form shells.

Horizontal broaching machines are applicable for both internal and external surfaces. Figure (a) and (b) shows the principle of operation of broaching for internal and external machining.

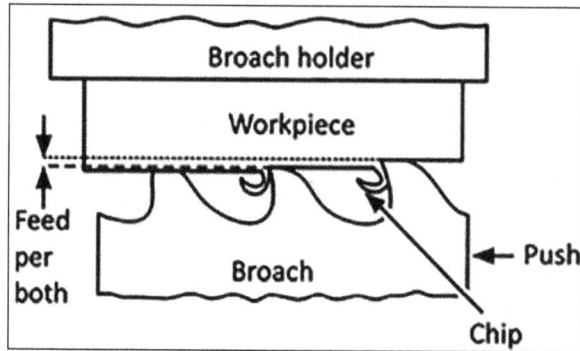

(a) A push type broach in use for machining an external surface.

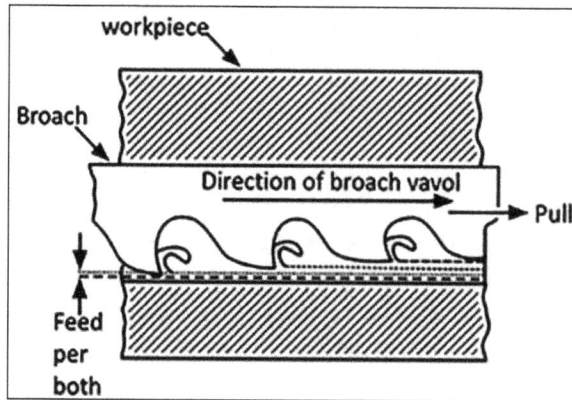

(b) A push type broach in use for machining an internal surface.

Figure (a) shows push type broach which is used for machining of external surface and figure (b) shows pull type broach which is used for machining of internal surface.

During the process, either a broach is kept stationary and work piece is fed or work piece is kept stationary and broach is fed. The surface produced by broaching carries an inverse profile as that of a broach used for operation.

Horizontal broaching machines have a bed similar to the lathe machine and the broach moves like a tail stock on the bed ways, Horizontal internal broaching machines range from 2 to 60 tones and stroke up to 3 m, whereas horizontal external or surface broaching are available up to 100 tones and stroke up to 9 m. Horizontal internal broaching are generally used for producing internal splines in the boss of a gear.

Surface Broaching Machine

- In surface broaching machine, either work piece or broach moves across each other.

- The process of broaching external work surface is known as surface broaching.

- These machines are generally vertical and hydraulically operated.

- Surface broaching machines is another alternative to milling machine and hence in these machines fixtures are also used.

- During operation, the broaching surface must be rigidly mounted and the work piece must be solidly supported as it passes the broaching section.

- These machines are used for large quantities of work and quite large, relatively the surfaces.

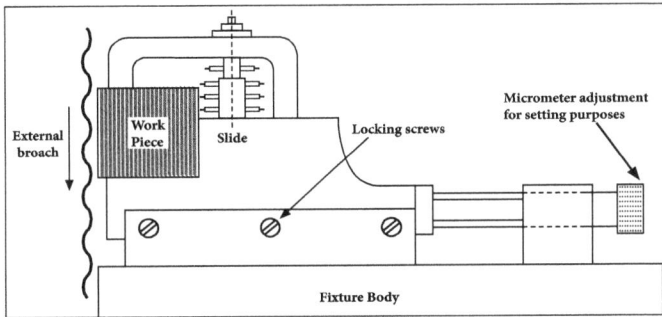

Surface Broaching.

Key way Broaching Machines

Construction and Working

- Figure shows the key way broach with its adapter which guides the broach as well as do the work of holding and locating the work piece.

- This is the simplest type of broaching machines and can be used for general purpose.

- If multiple key ways or splines are to be cut, a single broach can be used with the work piece and indexed after each cut.

- Broaching internal key ways is the oldest method.

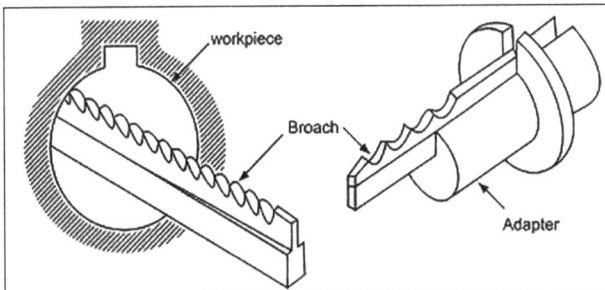

Key way Broaching Machines.

7.2   Types of Broaching Machines

The broaching machine may be classified as follows:

1) According to the nature and direction of primary cutting motion:

- Horizontal broaching machine.

- Vertical broaching machine.

- Continuous broaching machine.

2) According to the purpose:

- Internal broaching machine.

- External surface broaching machine.

3) According to method operation:

- Pull broaching machine.

- Push broaching machine.

Working of a Surface Broaching Machine

Broaching machines are probably the simplest of all machine tools. They consist of a work holding fixture, a broaching tool, a drive mechanism and a suitable supporting frame. Although the component parts are few, several variations in design are possible.

There are two principal types of machines:

- Horizontal

- Vertical.

In addition to these standard types. There are special and continuously operating machines. Both horizontal and vertical types have one or more rams depending on production requirement. Dual-ram models are arranged so that when one ram is on the cutting stroke, the other is on the returns stroke and the return stroke is performed quickly to gain time, which is used to unload and load the machine.

Broaching machines usually pull or push the broach through or past a work piece that is held in a fixture. On some machines, however, the work piece is moved past a broach that is fixed in its position. Most broaching machines are hydraulically operated to secure a smooth, uniform cutting action.

1) Horizontal broaching machines:

Nearly all horizontal machines are of the pull type. They may be used for either internal or external broaching, although internal work is the most common. A horizontal broaching machine shown in figure consists of a bed or a base a little more than twice the length of the broaching stroke, a broach pilot and the drive mechanism for pulling the broach.

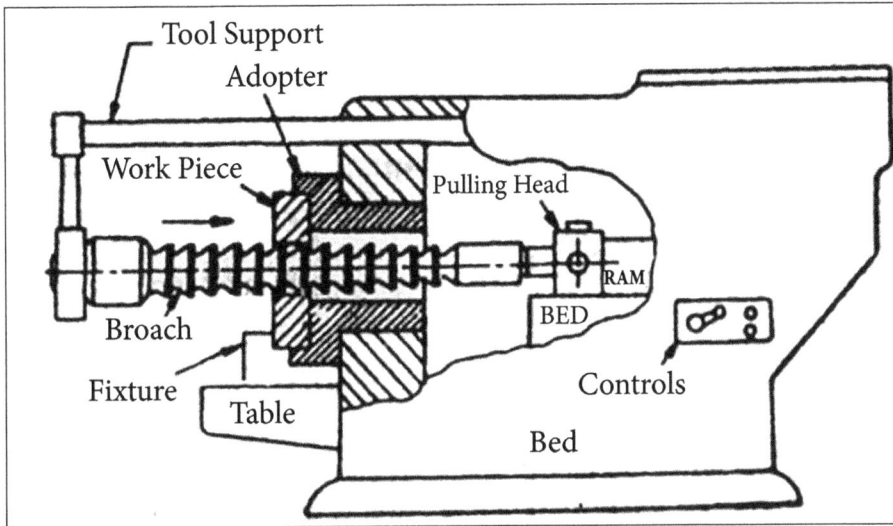

Horizontal broaching machine.

- Pulling head.

- Broach.

- Work fixture.

Horizontal broaching machines are used primarily for broaching key ways, splines, slots, round holes and other internal shapes or contours. They have the disadvantage of taking more floor space than do the vertical machines.

2) Vertical Broaching Machine:

The vertical broaching machine is used for large, heavy work parts through large diameters, frequently the work part diameter is more than its length. The part is clamped to worktable that rotates relative to machine base. Typical boring machine can place and feed several cutting tools at the same time.

Tools mounted on tool heads can be fed horizontally and vertically relative to worktable. One or two heads are mounted on horizontal cross-rail assemble to machine tool housing over the worktable. Cutting tools mount over the work can be used for lacing and boring. In adding to tools on cross-rail, one or two extra tool heads can mount on side columns of housing to enable revolving on outside diameter of work.

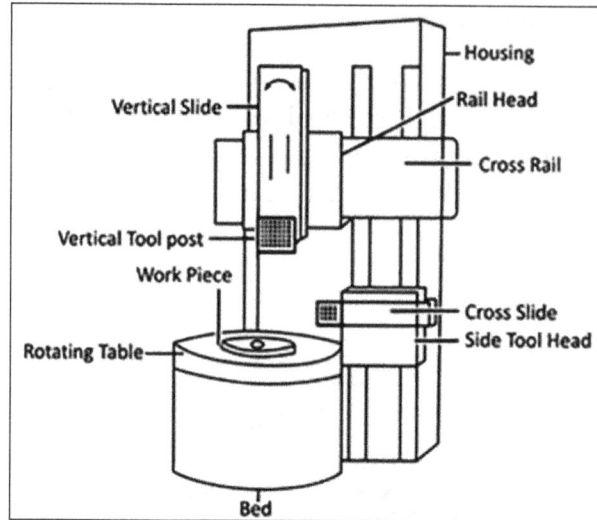

Vertical broaching machine.

Tool heads used on vertical boring machine frequently includes turrets to contain several cutting tools. This results in a loss of difference among this machine and a vertical turret lathe. A few machine tool builders create the distinction that the vertical turret lathe is used for work diameters up to 2.5 m (100 in), at the same time VAM is used for larger diameters.

Moreover, vertical boring mills are regularly applied to jobs, as vertical turret lathes are used for batch production.

Push Broaching Machines

In these machines the broach movement is guided by a ram. These machines are simple, since the broach only needs to be pushed through the component for cutting and then retracted. The work piece is fixed into a boring fixture on the table. Even simple arbor presses can be used for push broaching.

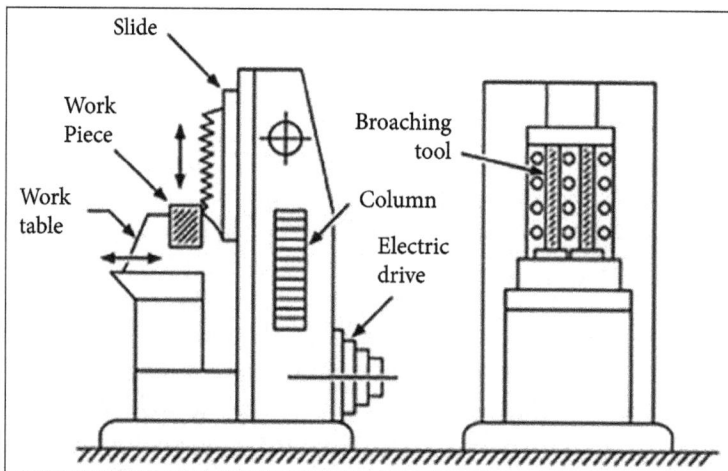

Push down type vertical surface broaching machine.

Above figure shows the push down type vertical surface broaching machine. It consists of a box shape column, slide and drive mechanism. Broach is mounted on the slide which is hydraulically operated and accurately guided on the column ways.

Slide with the broach travels at various speeds. The slide is provided with quick return mechanism. The worktable is mounted on the base in front of the column. The fixture is clamped to the table. The work piece is held in the fixture.

After advancing the table to the broaching position, it is clamped and the slide with the broach travel downwards for machining the work piece. Then the table recedes to load a new work piece and the slide returns to its upper position.

The same cycle is then repeated. Vertical broaching machines occupy less floor space and are more rigid as the ram is supported by the base. They are mostly used for external or surface broaching though internal broaching is also possible and occasionally done.

Pull Broaching Machines

These machines consist of a work holding mechanism and a broach pulling mechanism along with a broach elevator to help in the removal and threading of the broach through the work piece. The work piece is mounted in the broaching fixture and the broach is inserted through the hole present in the work piece.

Then the broach is pulled through the work piece completely and the work piece is then removed from the table. Afterwards the broach is brought back to the starting point before a new work piece is located on the table. The same cycle is then repeated.

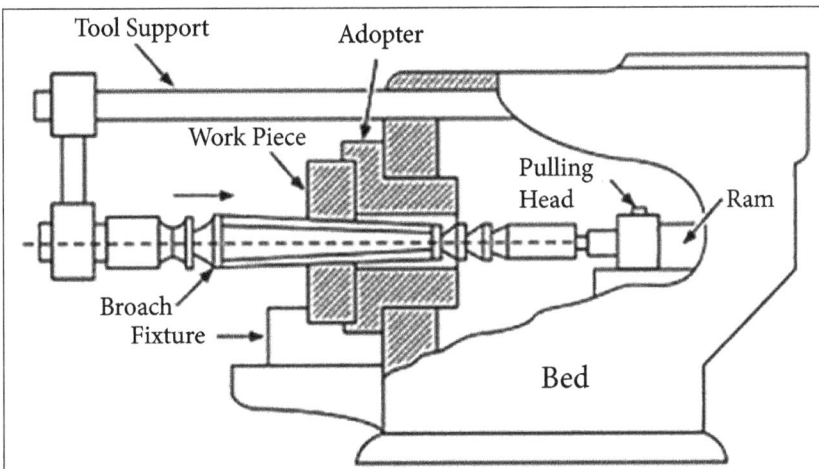

Pull type horizontal internal broaching machine.

Above figure shows the pull type horizontal internal broaching machine. This machine has a box type bed. The length of bed is twice the length of stroke. Most of the modern horizontal broaching machines are provided with hydraulic or electric drive. It is housed in the bed. The job is located in the adopter.

The adopter is fitted in the front vertical face of the machine. The small end of the broach is inserted through the hole of the job and connected to the pulling head. The pulling head is mounted in the front end of the ram.

The ram is connected to the hydraulic drive mechanism. The rear end of the broach is supported by a guide. The broach is moved along the guide ways. It is used for small and medium sized works. It is used for machining key ways, splines, serrations, internal gears, etc.

Horizontal broaching machines are the most versatile in application and performance and hence are most widely employed for various types of production. These are used for internal broaching but external broaching work is also possible. The horizontal broaching machines are usually hydraulically driven and occupy large floor space.

Pull down type vertical internal broaching machine:

- This machine has an elevator at the top. The pulling mechanism is enclosed in the base of the machine. The work piece is mounted on the table by means of fixture. The tail end of the broach is gripped in the elevator. The broach is lowered through the work piece.

- The broach is automatically engaged by the pulling mechanism and is pulled down through the job. After the operation is completed, the broach is raised and gripped by the elevator. The elevator returns to its initial position. This is illustrated in figure (a) below.

Pull up type vertical internal broaching machine:

- The ram slides on the vertical column of the machine. The ram carries the pulling head at its bottom. The pulling mechanism is above the worktable and the broach is in the base of the machine.

- The broach enters the job held against the underside of the table and is pulled upward. At the end of the operation, the work is free and falls down into a container. This is illustrated in figure (b) below.

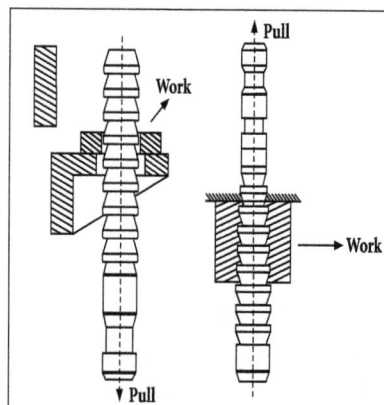

Vertical internal broaching operation (a) pull down type (b) pull up type.

Continuous Broaching Machine

There are three types of continuous broaching machines. They are horizontal, vertical and rotary types. These machines are used for mass production of small parts. These machines are used for surface broaching.

Horizontal Continuous Broaching Machine

A line diagram of this type is shown in the figure. This broaching machine has a driving unit. It consists of two sprocket. They are connected by an endless chain. A series of fixture arc mounted on this chain. They travel along with the chain in a fixed path. The broaches are fixed horizontally in the frame of the machine.

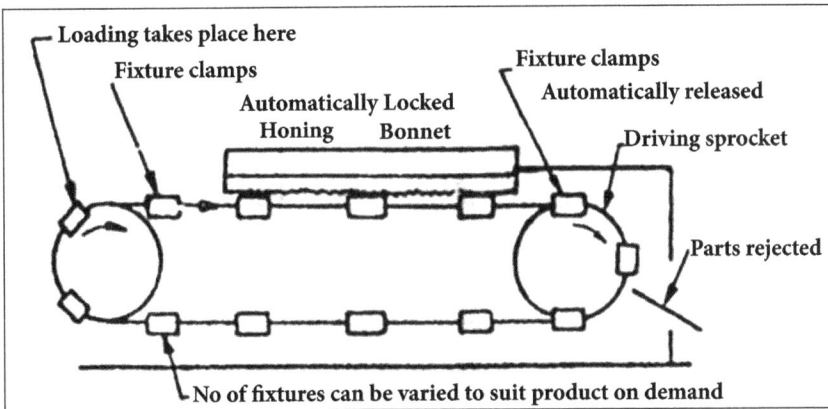

Horizontal type continuous broaching machine.

When the fixture passes the loading station, the operator drops the parts in the fixture. The work is automatically clamped before it reaches the tunnel. The work pieces move under the broach. Then broaching takes place. The work pieces are automatically released by a cam. At the unloading point the work pieces fall out of the fixtures.

Rotary Table Continuous Broaching Machine

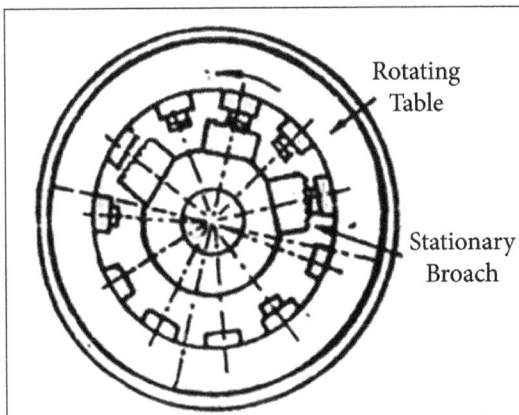

Rotary table continuous broaching machine.

The machine has a rotary table and a vertical column. In the vertical column, the broach is fitted horizontally above the table. Fixtures are mounted on the rotary table. The work pieces are clamped on these fixtures. They move past the stationary broaches.

Rotary broaching machines are limited to small parts. They are used for squaring distributor shaft, slotting and the facing of small parts. Figure shows one of the types of rotary table continuous broaching machine.

It consists of a rotary table. Parts are fixed on the inner periphery of the rotating table. The broaches are arranged on the central stationary part.

Progressive Broach

In this broach, some of its teeth at front end have the same height but have different width. The last teeth of this broach remove less material over the entire profile. The roughing teeth of this broach remove more material per tooth.

The cutting edges of the roughing teeth cut the metal not along the width of the broached surface but in narrow strips. The finishing teeth finish off the work surface along its full width. It is shown in the figure.

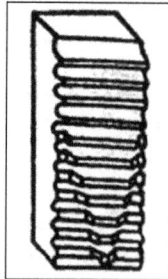

Progressive broach.

Solid Broaching

When a broach is made as a single piece, it is known as solid broach. Broaches for internal breaching are solid type. But solid broaches for surface broaching is also available as shown in the figure.

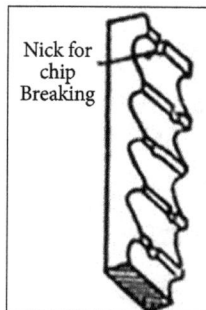

Solid broaching.

Various Operations Performed by Broaching Machine

Broaching is applied for machining a key way in a hole, as internal/external or round/contoured and surface or a spline hole etc. Some of the important broaching operations are given below:

Broaching Splines

A hole having number of slots is called splines. Straight-sided splines can be broached by the normal key way broaching procedure. The broach has four or more rows of teeth on its periphery according to the splines to be produced.

The work is clamped on a suitable fixture. The broach is inserted in the hole of the work. Then it is pulled through the work. This operation is performed by internal broaching machine.

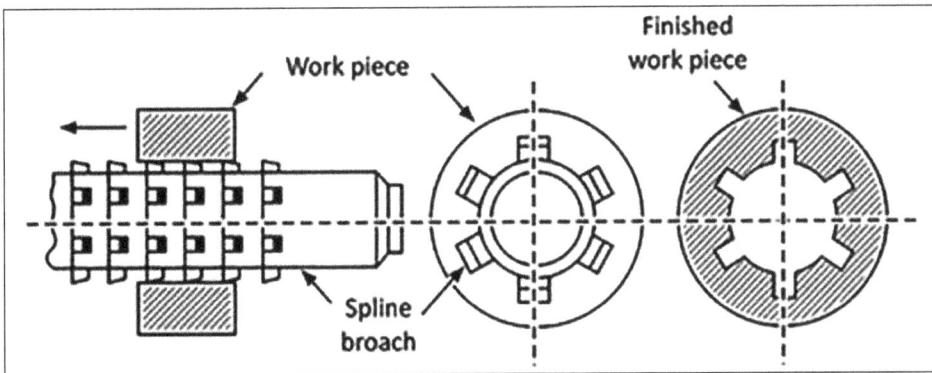

Spline Broaching.

Broaching a Key way

Broaching of internal key way can be done easily on internal broaching machines. There is a standard guide bushing with a rectangular slot to support and guide the broach.

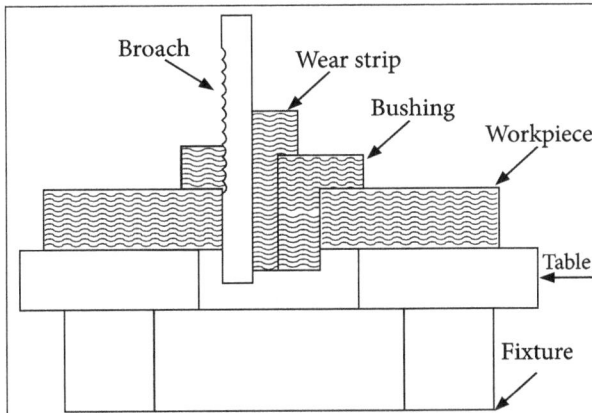

Key way broaching.

Bushing is placed in the bore of the work and the front pilot of the broach is pushed in the slot until the first tooth gets in contact with the edge of the work. The required force is applied to push the broach through the work. Several passes are required to machine the required depth.

Applications:

- Special gear manufacturing.

- Compressor wheels manufacturing.

- Bushings and sleeves manufacturing.

- Turbine blades, rotors, etc.

7.2.1 Advantages and Limitations

Advantages:

- It gives high rate of production so recommended for mass production.

- Production run time in case of broaching is very large as a broach has very long life. The whole processing load is shared by so many teeth.

- Due to faster operation and longer tool life, it is relatively cheaper.

- Both rough cutting as well as finished cutting can be completed in a single pairs of tool.

- Little skill or employment of semi-skilled manpower is sufficient to perform broaching operation.

- Broaching provides accurate and excellent quality of surface finish. It is capable to maintain tolerance of the order of 0.007 mm and surface finish CLA value up to 0.8.

- It is also capable to process internal and external surface including intricate shaped cavities.

- Broaching makes the effective use of cutting fluids as it facilitates the flow of cutting fluid into the cuts.

Limitations:

- Broach is a multi-point cutting tool having multi cutting edges. Preparation of cutting edges is a costly affair. Its initial cost is quite high.

- There is a limitation of size of work piece in case of broaching. Very large sized work pieces cannot be subjected to broaching operation.

- Broaching is not possible for the surfaces having obstructions.

- Application of broaching is restricted up to finishing and accurate sizing as it can remove only small stocks of material. Removal of larger stocks is not possible in broaching operation.

- There is a urgent need of rigid clamping of work piece in broaching operation to maintain its accuracy and finish. Clamping devices require frequent maintenance and cost.

7.3   Finishing and other Processes

Honing Process

- Honing is an abrading process, used for finishing already machined surfaces.

- Mostly, honing is used for finishing internal cylindrical surfaces such as drilled holes.

- The tool used during the process is called as hone which is bonded with abrasive stones and made in the form of a stick.

- Honing can be done by hand or by using machines.

- The honing stones are held in holder or mandrel and forced outwards by hydraulic or mechanical pressure.

- To form the honing stones aluminum oxide, silicon carbide or diamond grams of suitable grit are bonded in resinoid, vitrified or shellac bond.

- In some cases vertical honing machines are also used.

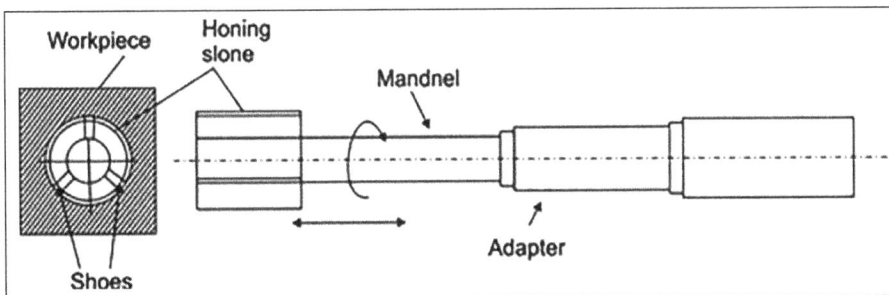

Hand honing tool and honing process.

- Figure shows the honing tool used in vertical honing machines.

- Out of all the surface finishing operations, honing enables maximum stock removal, still it is not a metal removing process.

- Due to this ability, honing is used for correcting slight out of roundness or taper.

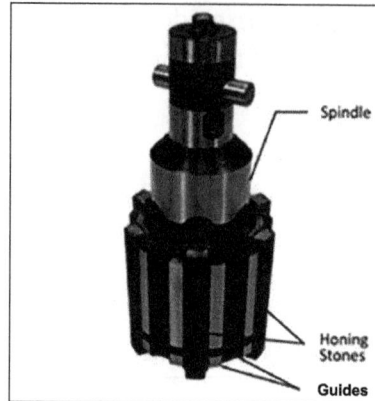

Honing tool head for vertical machines.

- The general amount of tool removed by honing process is from 0.1 mm to 0.25 mm, though it is capable of removing the stock up to 0.75 mm.

- The speed used for honing process is in the range of 10 to 30m/min.

Advantages of Honing:

- Desired surface finish can be obtained.

- The size can be controlled accurately.

- At low cost, high productivity is obtained.

- Honing can be done on most materials from aluminum to brass to hardened steel. Also, carbides, ceramics and glass can be honed with the help of diamond stones.

- As work piece need not be rotated by power, no chucks face plates or rotating tables are required.

- Rapid and economical stock removal with minimum heat dissipation and distortion is achieved.

Disadvantages of Honing:

- It is impossible to improve lack of straightness of holes.

- Horizontal honing may create oval holes unless the work piece is rotated or supported.

Applications of Honing:

- Mainly, honing is used for removing scratches left-over after grinding.

- Also used for removing stocks from internal and external surfaces.

Lapping

This operation employs a lapping plate made of east iron (the porosity helps to lodge the abrasive grains) or some softer material. During lapping, the fine abrasive which is mixed with grease and called lapping compound is smeared and rubbed under pressure against the hard work piece surface. This causes erosion of the harder work piece.

The softer lap is now worn out during lapping because the abrasive grain gets embedded in the softer lap instead of cutting it. It is important that the lapping plate follows a non-repeating path such as figures of eight to avoid formation of directional scratch pattern. Lapping gives high degree of flatness or roundness depending on the geometry of the lap.

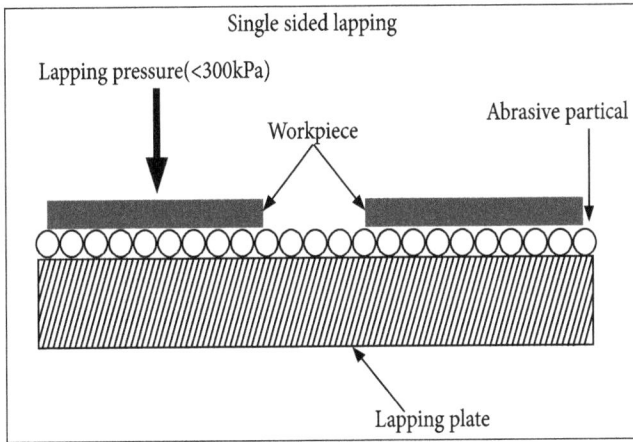

Lapping process.

Lapping Operations

1. Flat planetary lapping: The figure shows the vertical flat lapping process where large quantities of similar parts are being handled and, in some cases, both surfaces are machined simultaneously. The resultant parallelism and the uniformity of dimensions are better than that of hand lapping.

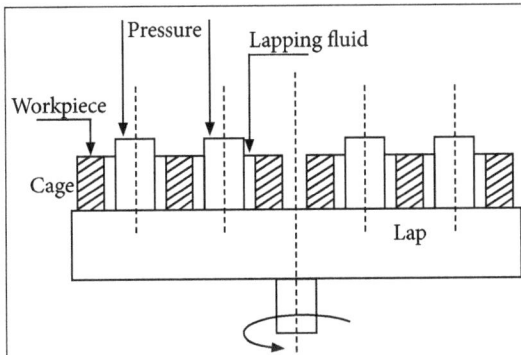

Flat planetary lapping.

2. Spherical lapping: The figure shows the lapping of spherical surfaces other than balls. Accordingly, the lap is a counterpart of the surface to be machined. The lap should be heavy enough to provide the required pressure.

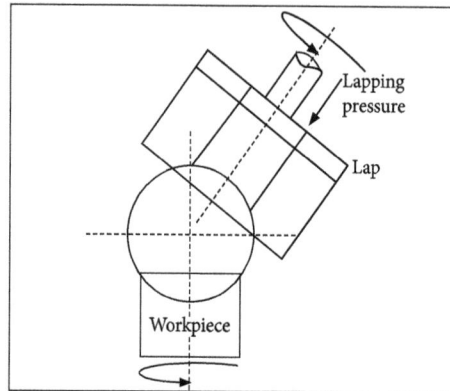

Spherical lapping.

3. Vibratory lapping: To increase the linear material removal rate by lapping, additional vibration is applied to the lap as shown in the figure. Under such conditions, the material removal rates rise by 30%-40%, but the height of surface irregularities increases by 50%-100%.

Vibratory lapping is, therefore, suitable as a preliminary lapping process or when the surface required is not smooth. The abrasive mixture of boron carbide or diamond dust is used for longer abrasive life and material removal rate requirements.

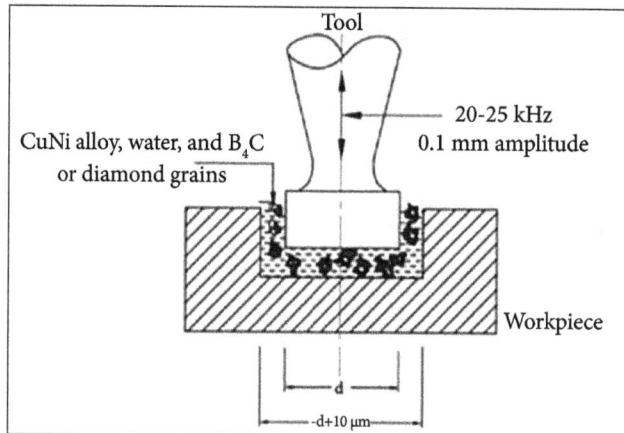

Vibratory lapping.

7.3.1 Super Finishing Process

1. Honing:

It is mostly used for finishing internal cylindrical surfaces such as drilled or bored holes.

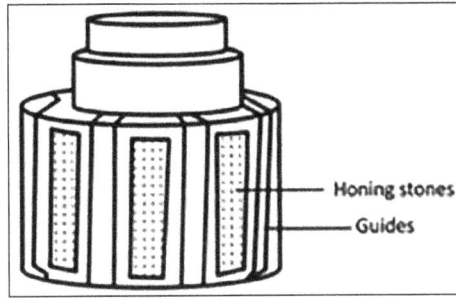
Honing.

2. Super Finishing:

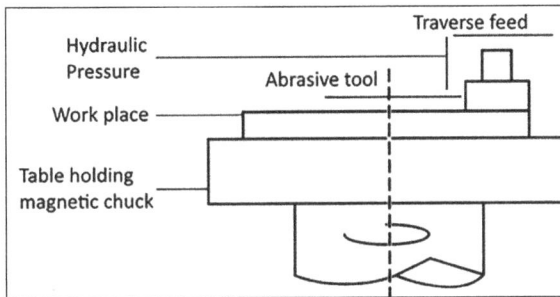
Super finishing the flat surface.

It is an abrading process, efficient in surface refining of cylindrical, flat, spherical and cone shaped parts.

3. Lapping:

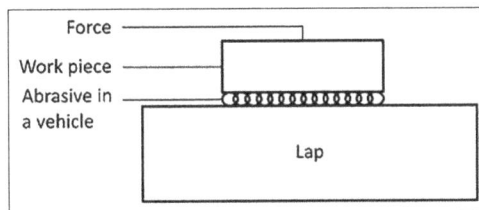
Lapping.

Lapping is a finishing process carried out after grinding and designed to produce an exceptionally high degree of surface finish as well as a perfectly true surface accurate to size within extremely close limits.

Super finishing is the abrasive finishing process in which the working motions include:

- Oscillatory motion of the tool, i.e., reciprocating motion of short stroke and high frequency in the direction parallel to the axis of work piece rotation.

- Rotary motion of the work piece.

- Feed motion of the tool or work piece.

In straight super-finishing as shown in the figure (a), the feed motion is parallel to the work piece axis, in radial super-finishing as shown in the figure (b), it is perpendicular to that axis. In plunge super-finishing as shown in the figure (c), there is no stick (tool) feed, while in the internal super-finishing shown in the figure (d), the tool feed direction is axial.

The aim of oscillatory super-finishing is not to correct the shape and dimensional accuracy but to improve the surface finish and the quality of the surface layer. The super-finishing allowances are smaller than that in honing. They are equal to the mean total height of surface irregularities resulting from preliminary machining plus a part of the surface layer damaged in the latter operation.

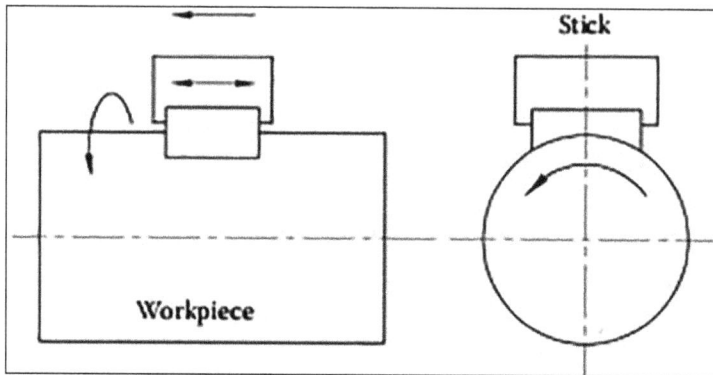

(a) Straight oscillatory super-finishing.

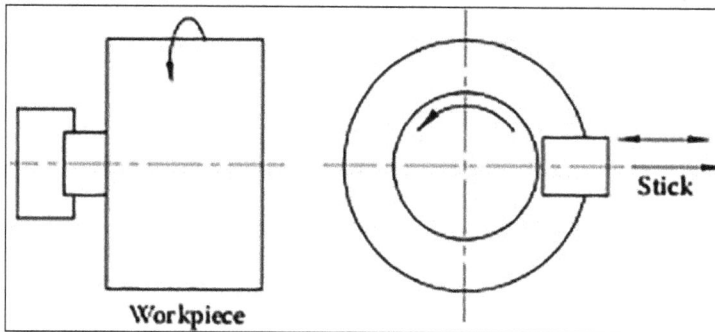

(b) Radial oscillatory super-finishing.

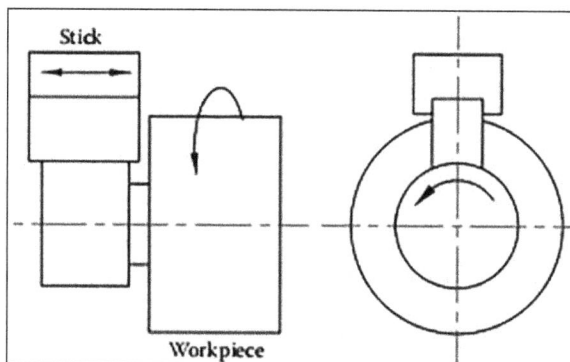

(c) Plunge oscillatory super-finishing.

Thus, the super-finishing allowances are often contained within the limits of dimensional accuracy. The rough-ness obtainable is 0.01 μm R_a, which offers high wear resistance and a high load-carrying capacity as compared to ground and precision-turned surfaces.

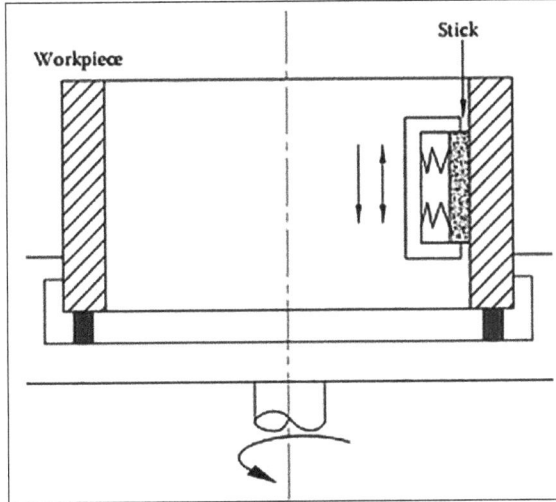

(d) Internal oscillatory super-finishing.

Super-finishing is efficient for finishing cylindrical, flat, spherical and conical surfaces. Although it is not suitable for changing dimensions, an average stock of 0.005–0.030 mm in diameter can be removed.

The high degree of surface improvement is achieved at a lower cost, compared to other finishing methods, because only a short time is required to obtain the finish and there are only a small percentage of rejected components.

As in the lapping process, the high quality of surface obtained in super-finishing is mainly due to:

- Low specific pressure of the abrasive stone.

- Low cutting speeds.

- Oscillation of the abrasive sticks.

- Low temperature generated (1°C–28°C above ambient).

- Combination of oscillation and traverse motions brings new grains in contact with work piece and facilitates the removal of chips away from the machining zone by the coolant.

Super-finishing is applied for external and internal surfaces of cast iron, steel and non-ferrous parts, which have been previously ground or precision turned. It is capable of rendering a high-quality surface finish.

7.3.2 Polishing, Buffing Operation and Application

Polishing

Polishing is the smoothing of surfaces by the cutting action of abrasive particles adhered to the surface of resilient wheels of wood, felt, leather and canvas or fabric or attached to belts operating on resilient wheels. The process is used to impart high grade of surface finish for the sake of appearance.

It is not used to control part size. Artificial abrasives like Al_2O_3 and SiC are commonly used. Flint, emery and garnet are used as natural abrasives. The mesh size ranges from 12 to 400. For best results in polishing, the wheels or belts should run at 3000 m/min.

Buffing Operation and Application

Buffing is a polishing operation in which the work piece is brought in contact with a revolving cloth buffing wheel, that usually has been charged with a very fine abrasive an figure. The polishing action in buffing is very closely related to lapping in that when a polishing medium such as `rouge' is used, the cloth buffing wheel becomes a carrying vehicle for the fine abrasives.

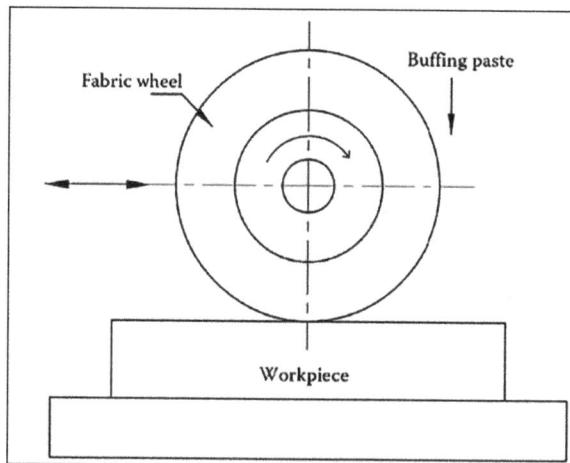

Buffing operation.

In this action the abrasive removes amounts of metal from the work piece, thus eliminating the scratch marks and producing a very smooth surface. When softer metals are buffed, particularly without the use of an abrasive, there is some indication that a small amount of metal flow may occur which helps to reduce the high spots and produce a high polish.

Buffing wheels are made of discs of linen, cotton, broad cloth and canvass. They are made more or less firm by the amount of stitching used to fasten the layers of the cloth together. Buffing wheels for very soft polishing or which can be used to polish into interior corners may have no stitching, the cloth layers being kept in position by the centrifugal force resulting from the rotation of the wheel. Buffing wheel speeds are in the range of 32.5 to 40 m/s.

Various types of huffing rouges are available. Most of them being primarily ferric oxide in some soft type of binder. Buffing should be used only to remove very fine scratches or to remove oxide or similar coatings which may he on the work surface. It ordinarily, is done manually, die work being held against the rotating wheel.

This procedure is apt to be relative expensive because of the labour cost. There are semi-automatic buffing machines available consisting of a series of individually driven buffing wheel which can be adjusted to the desired position so as to buff different portions of the work piece. The work pieces are held in fixtures on a rotating circular worktable so as to move past the bulling wheels.

If the work pieces are not too complex in shape, very satisfactory results can be achieved with such equipment and the buffing cost will be low. Product applications of buffing process which produces mirror-like finish are objects used on mobile homes, automobiles, motor-cycles, boats, bicycles, sporting items, tools, store fixtures, commercial and residential hardware and household utensils and appliances.

Comparison between Polishing and Buffing Processes

	Polishing	Buffing
1	It is employed for removing scratches, tool marks and other irregularities from the work piece surface which are produced through other applications.	It is performed after polishing for providing a high luster to the polished surface.
2	It is performed by using abrasive coated wheels or belts.	The wheels used in buffing process are generally made of cloth (wool, cotton or muslin) or fiber which is charged with loose abrasive grains.
3	Can maintain flatness or roundness.	It does not maintain flatness or roundness.

Non-Traditional Machining Processes

8.1   Need for Non-Traditional Machining

Conventional machining processes utilize the ability of the cutting tool to stress the material beyond the yield point to start the material removal process. This requires that the cutting tool material is harder than the work piece material. New materials which are having high strength to weight ratio, heat resistance and hardness, such as nimonic alloys, alloys with alloying elements such as tungsten. Molybdenum, columbiums are difficult to machine by the traditional methods.

Machining of these materials by the conventional methods is very difficult as well as time consuming, since the material removal rate reduces with an increase in the work material hardness. Hence, there is the need for development of non-traditional machining processes which utilize other methods such as electro-chemical processes for the material removal. As a result, these processes are termed as unconventional or non-traditional machining methods.

The complex shapes in these materials are either difficult to machine or time consuming by the traditional methods. In such cases, the application of the non-traditional machining processes finds extensive use. Further, in some applications a very high accuracy is desired besides the complexity of the surface to be machined.

These processes are not meant for replacing the conventional processes but to supplement them.

There are a number of processes available in these. They are:

- Electric Discharge Machining (EDM).

- Electro Chemical Machining (ECM).

- Electro Chemical Grinding (ECG).

- Ultrasonic Machining (USM).

- Laser Beam Machining (LBM).

- Chemical machining (CHM).

- Abrasive Water Jet Machining (AWJM).

- Water Jet Machining (WJM).

- Plasma Arc Machining (PAM).

The conventional metal cutting processes make use of the shearing process as the basis for material removal.

The Major Characteristics of Non-conventional Machining

Non-conventional manufacturing processes is defined as a group of processes that remove excess material by various techniques involving mechanical, thermal, electrical or chemical energy or combinations of these energies but do not use a sharp cutting tools as it needs to be used for traditional manufacturing processes.

The major characteristics of Non-conventional machining are:

- Material removal may occur with chip formation or even no chip formation may take place. For example in AJM, chips are of microscopic size and in case of Electrochemical machining material removal occurs due to electrochemical dissolution at atomic level.

- In NTM, there may not be a physical tool present. For example in laser jet machining, machining is carried out by laser beam. However in Electrochemical Machining there is a physical tool that is very much required for machining.

- In NTM, the tool need not be harder than the work piece material. For example, in EDM, copper is used as the tool material to machine hardened steels.

- Mostly NTM processes do not necessarily use mechanical energy to provide material removal. They use different energy domains to provide machining.

8.1.1 Operation Laser Beam Machining

As the name implies in Laser beam Machining the source of energy is the LASER (Light Amplification by stimulated Emission of Radiation). The laser beam focuses optical energy on the surface of the work piece.

A laser beam can be so powerful when used with lens system that it can melt and vaporize diamond as the energy density can be of the order of 105 kW/cm^2. This huge amount of energy is released due to some specific atoms having higher energy levels and particular frequency.

Laser Beam Machining Principles with Schematics

Different types of lasers are used in Laser beam machining (LBM).

For example – solid state, gas and semiconductor.

At times high power lasers are required for machining and welding and in those cases only solid state lasers can provide such power levels.

Ruby-laser or crystalline aluminium oxide or sapphire is the most commonly used solid state laser.

Generally these lasers are fabricated in into rods having length about 150 mm. Their ends are well furnished to close optical tolerances. Figure below shows a schematic view of laser beam machining process.

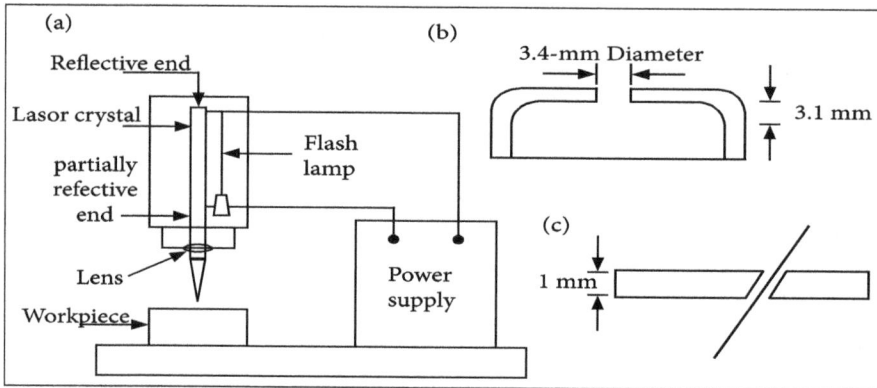

Schematic illustration of the laser-beam machining process.

A small amount of chromium oxide is added to dope the ruby crystal. A flash of high intensity light, generally Xenon-filled flash lamp is used to pump the laser. To fire the xenon lamp a large capacitor is required to be discharged through it and 250 to 1000 watts of electric power is needed to do this.

The intense radiation discharged from the lamp excites the fluorescent impurity atoms (chromium atoms) and these atoms reaches a higher energy level. After passing through a series of energy levels when the atoms fall back to original energy level, an intense beam of visible light emission is observed.

This beam is reflected back from the coated rod ends and make more and more atoms excited and stimulated and return to ground level. A stimulated avalanche of light is obtained which is transmitted through the coated part (~80% reflective). This light which is highly coherent in time and space has a very narrow frequency band, is highly in phase and quite parallel.

If this light is focused in association with ordinary lenses on the desired spot of the w/p, high energy density is gained which helps to melt and vaporize the metal.

Advantages of Laser Beam Machining:

- No tool wear as there is no direct contact between tool and work piece.

- Metal and non-metals (e.g. plastics and rubbers) irrespective of their brittleness and hardness can be machined.

- Laser beam can go through a long distance as a result LBM can be used to weld, drill or cut areas which are difficult to reach.

- Laser beam welding gives the opportunities to weld/cut magnetic as well as heat treated materials without losing their properties. (some change in the properties is observed in the heat affected zone).

- Any environment is suitable for laser beam machining – through transparent medium and magnetic fields.

- Very little distortion is observed and tow materials can be easily joined together. (Vaporization of the metal is not expected so it must be avoided).

- Difficult-to-machine or refractory materials can be drilled.

- Micro sized holes can be created in all types of materials.

- Energy obtained is of high density as a result high heat is obtained.

- Beam configuration and size of exposed area is easily controllable.

- Precise location of the spot is ensured.

- By applying unidirectional multiple pulses deep holes of very short diameter can be drilled.

Disadvantages of LBM:

- The initial cost is very high and lifespan of the flash lamp is short.

- The safety procedures are needed to be followed very strictly.

- Material removal rate is not up to the mark.

- While machining some plastics burn or char is noticed.

- Too deep holes are not possible to drill.

- Machined holes are not round shaped or straight.

- Overall efficiency is very low. (0.3~0.5 %).

Applications:

- Welding of non-conductive and refractory material.

- Cutting complex profiles for both thin and hard materials.

- Used to make tiny holes. Such as the holes in the nipples of baby feeder.

- Mass-micro machining.

- Can be used for dynamic balance of rotating parts.

- Some special heat treatment of materials.

- For producing fine and minute holes.

8.1.2 Plasma Arc Machining

It is also one of the thermal machining processes. Here the method of heat generation is different than EDM and LBM.

Working Principle of PAM

In this process gases are heated and charged to plasma state. Plasma state is the superheated and electrically ionized gases at approximately 5000° C. These gases are directed on the work piece in the form of high velocity stream. Working principle and process details are shown in Figure.

Process Details of PAM

Plasma Gun

Gases are used to create plasma like, nitrogen, argon, hydrogen or mixture of these gases. The plasma gun consists of a tungsten electrode fitted in the chamber.

The electrode is given negative polarity and nozzle of the gun is given positive polarity. Supply of gases is maintained into the gun.

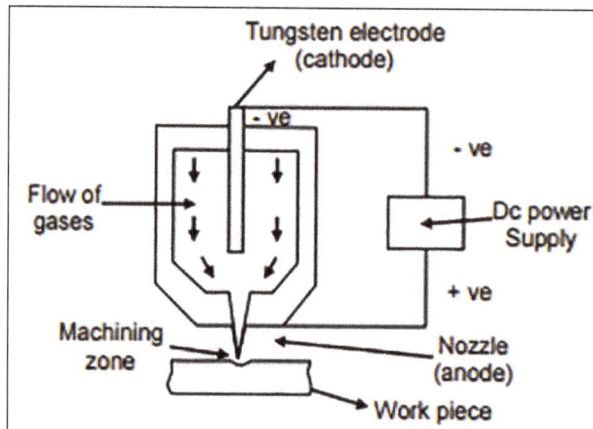

A strong arc is established between the two terminals anode and cathode. There is a collision between molecules of gas and electrons of the established arc.

As a result of this collision, gas molecules get ionized and heat is evolved. This hot and ionized gas called plasma is directed to the work piece with high velocity. The established arc is controlled by the supply rate of gases.

Working Principle and Process Details of PAM

Power Supply and Terminals

Power supply (DC) is used to develop two terminals in the plasma gun. A tungsten electrode is inserted to the gun and made cathode and nozzle of the gun is made anode. Heavy potential difference is applied across the electrodes to develop plasma state of gases.

Cooling Mechanism

As we know that hot gases continuously comes out of nozzle so there are chances of its overheating. A water jacket is used to surround the nozzle to avoid its overheating.

Tooling

There is no direct visible tool used in PAM. Focused spray of hot, plasma state gases works as a cutting tool.

Workpiece

Workpiece of different materials can be processed by PAM process. These materials are aluminium, magnesium, stainless steels and carbon and alloy steels. All those material which can be processed by LBM can also be processed by PAM process.

Applications of PAM

The chief application of this process is profile cutting as controlling movement of spray focus point is easy in case of PAM process. This is also recommended for smaller machining of difficult to machining materials.

Advantages of PAM Process:

- It gives faster production rate.

- Very hard and brittle metals can be machined.

- Small cavities can be machined with good dimensional accuracy.

Disadvantages of PAM Process:

- Its initial cost is very high.

- The process requires over safety precautions which further enhance the initial cost of the setup.

- Some of the work piece materials are very much prone to metallurgical changes on excessive heating so this fact imposes limitations to this process.

- It is uneconomical for bigger cavities to be machined.

8.1.3 Electro Chemical Machining

Electrochemical machining (ECM) process uses electrical energy in combination with chemical energy to remove the material of work piece. This works on the principle of reverse of electroplating.

Working Principle of ECM

Electrochemical machining removes material of electrically conductor work piece. The work piece is made anode of the setup and material is removed by anodic dissolution. Tool is made cathode and kept in close proximity to the work piece and current is passed through the circuit. Both electrodes are immersed into the electrolyte solution. The working principle and process details are shown in the figure. This works on the basis of Faraday's law of electrolysis. The cavity machined is the mirror image of the tool. MRR in this process can easily be calculated according to Faraday's law.

Workpiece

Workpiece is made anode, electrolyte is pumped between work piece and the tool. Material of work piece is removed by anodic dissolution. Only electrically conducting materials can be processed by ECM.

Tool

A specially designed and shaped tool is used for ECM, which forms cathode in the ECM setup. The tool is usually made of copper, brass, stainless steel and it is a mirror image of the desired machined cavity. Proper allowances are given in the tool size to get the dimensional accuracy of the machined surface.

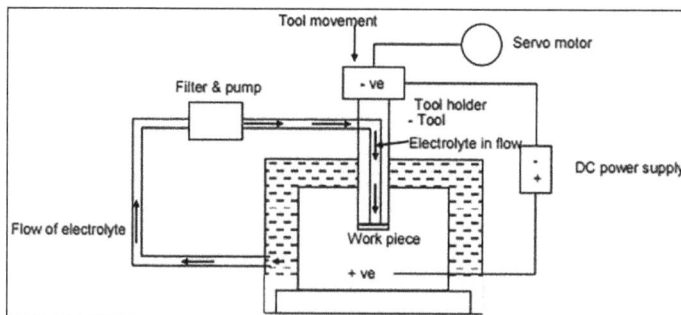

Working Principle and Process Details of ECM

Power Supply

DC power source should be used to supply the current. Tool is connected with the negative terminal and work piece with the positive terminal of the power source. Power supply supplies low voltage (3 to 4 volts) and high current to the circuit.

Electrolyte

Water is used as base of electrolyte in ECM. Normally water soluble NaCl and NaNO3 are used as electrolyte. Electrolyte facilitates are carrier of dissolved work piece material. It is recycled by a pump after filtration.

Tool Feed Mechanism

Servo motor is used to feed the tool to the machining zone. It is necessary to maintain a constant gap between the work piece and tool so tool feed rate is kept accordingly while machining.

In addition to the above, whole process is carried out in a tank filled with electrolyte. The tank is made of transparent plastic which should be non-reactive to the electrolyte. Connecting wires are required to connect electrodes to the power supply.

Applications of ECM Process

There is large number of applications of ECM some other related machining and finishing processes as described below:

- Electrochemical Grinding: This can also be named as electrochemical deburring. This is used for anodic dissolution of burrs or roughness a surface to make it smooth. Any conducting material can be machined by this process. The quality of finish largely depends on the quality of finish of the tool.

- This is applied in internal finishing of surgical needles and also for their sharpening.

- Machining of hard, brittle, heat resistant materials without any problem.

- Drilling of small and deeper holes with very good quality of internal surface finish.

- Machining of cavities and holes of complicated and irregular shapes.

- It is used for making inclined and blind holes and finishing of conventionally machined surfaces.

Advantages of ECM Process:

- Machining of hard and brittle material is possible with good quality of surface finish and dimensional accuracy.

- Complex shapes can also be easily machined.

- There is almost negligible tool wear so cost of tool making is only one time investment for mass production.

- There is no application of force, no direct contact between tool and work and no application of heat so there is no scope of mechanical and thermal residual stresses in the workpiece.

- Very close tolerances can be obtained.

Disadvantages and Limitations of ECM:

- All electricity non-conducting materials cannot be machined.

- Total material and work piece material should be chemically stable with the electrolyte solution.

- Designing and making tool is difficult but its life is long so recommended only for mass production.

- Accurate feed rate of tool is required to be maintained.

8.1.4 Ultrasonic Machining

Ultrasonic machining (USM) is one of the non-traditional machining process. Working principle of this process resembles with conventional and metal cutting as in this process abrasives contained in a slurry are driven at high velocity against the work piece by a tool vibrating at low amplitude and high frequency.

Amplitude is kept of the order of 0.07 mm and frequency is maintained at approximately 20,000 Hz. The work piece material is removed in the form of extremely small chips.

Normally very hard particle dust is included in the slurry like, Al2O2, silicon carbide, boron carbide or diamond dust. Working principle of USM is same as that of conventional machining that is material of work piece is removed by continuous abrasive action of hard particles vibrating in the slurry.

Abrasive slurry acts as a multipoint cutting tool and does the similar action as done by a cutting edge.

Process Details

USM process is indicated in line diagram shown in Figure. Details of the process are discussed below.

Abrasive Slurry

Abrasive slurry consists of dust of very hard particles. It is filled into the machining zone. Abrasive slurry can be recycled with the help of pump.

Work Piece

Work piece of hard and brittle material can be machined by USM. Work piece is clamped on the fixture 'I' the setup.

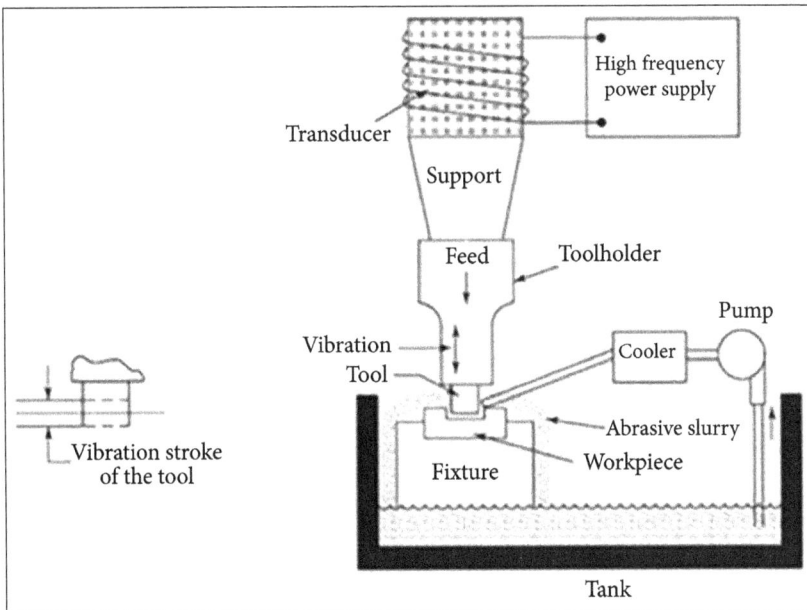

Details of USM Process.

Cutting Tool

Tool of USM does not do the cutting directly but it vibrates with small amplitude and high frequency. So it is suitable to name the tool as vibrating tool rather than cutting tool. The tool is made of relatively soft material and used to vibrate abrasive slurry to cut the work piece material. The tool is attached to the arbor (tool holder) by brazing or mechanical means. Sometimes hollow tools are also used which feed the slurry focusing machining zone.

Ultrasonic Oscillator

This operation uses high frequency electric current which passes to an ultrasonic oscillator and ultrasonic transducer. The function of the transducer is to convert electric energy into mechanical energy developing vibrations into the tool.

Feed Mechanism

Tool is fed to the machining zone of work piece. The tool is shaped as same to the cavity to be produced into the work piece. The tool is fed to the machining area. The feed rate is maintained equal to the rate of enlargement of the cavity to be produced.

Applications of USM

This process is generally applied for the machining of hard and brittle materials like carbides glass, ceramics, precious stones, titanium, etc. It is used for tool making punch and die making. The work piece material is normally removed in the form of very fine chips, so generated surface quality is extremely good. It is widely used for several machining operations like turning, grinding, trepanning and milling, etc. It can make hole of round shape and other shapes.

Advantages of USM:

- Its main advantage is the work piece after machining is free from any residual stress as to concentrated force or heat is subject to it during the machining process.

- Extremely hard and brittle materials can be machined, their machining is very difficult by conventional methods.

- Very good dimensional accuracy and surface finish can be obtained.

- Operational cost is low.

- The process is environmental friendly as it is noiseless and no chemical and heating is used.

Disadvantages of USM:

- Its metal removal rate (MRR) is very low and it cannot be used for large machining cavities.

- Its initial setup cost and cost of tool is very high, frequency tool replacement is required as tool wear takes place in this operation.

- Not recommended for soft and ductile material due to their ductility.

- Power consumption is quite high.

- Slurry may have to be replaced frequently.

8.2   Abrasive Jet Machining (AJM)

Equipment

In AJM, air is compressed in an air compressor and compressed air at a pressure of around 5 bar is used as the carrier gas as shown in the figure, also shows the other major parts of the AJM system. Gases like CO_2, N_2 can also be used as carrier gas which may directly be issued from a gas cylinder. Generally oxygen is not used as a carrier gas first passed through a pressure regulator to obtain the desired working pressure.

The gas is then passed through an air dryer to remove any residual water vapour. To remove any oil vapour or particulate contaminant the same is passed through a series of filters. Then the carrier gas enters a closed chamber known as the mixing chamber. The abrasive particles enter the chamber from a hopper through a metallic sieve.

The sieve is constantly vibrated by an electromagnetic shaker. The mass flow rate of abrasive (15 gm/min) entering the chamber depends on the amplitude of vibration of the sieve and its frequency. The abrasive particles are then carried by the carrier gas to the machining chamber via an electromagnetic on-off valve.

The machining enclosure is essential to contain the abrasive and machined particles in a safe and ecofriendly manner. The machining is carried out as high velocity (200 m/s) abrasive particles are issued from the nozzle onto a work piece traversing under the jet.

Abrasive Jet Machining Working

This abrasive and gas mixture emerge from a small nozzle mounted on a fixture at high velocity ranging from 150 to 300 m/min.

The Abrasive powder feed rate is controlled by the amplitude of vibration of the mixing chamber. A pressure regular controls the gas flow and pressure.

To control the size and shape of the cut either the work piece or nozzle is moved by Cams, pantographs, mechanisms. The carrier gas should be cheap, non-toxic and easily available. Air and nitrogen are two of the most widely used gas in AJM.

The Abrasives generally employed are aluminum oxide, silicon carbide, grass powder. The average particle size vary from 10 to 50 NS, larger sizes are used for rapid removal rate while smaller sizes are used for good surface finish and precision work.

Since nozzles are subjected to a great degree of abrasion wear, they are made of hard materials such as tungsten carbide or synthetic sapphire to reduce the wear rate.

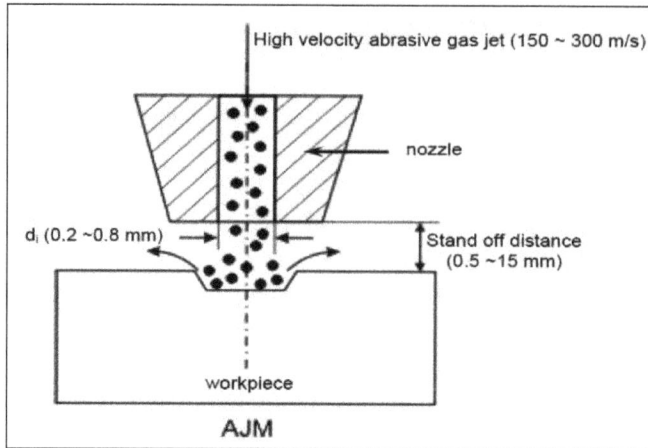

Abrasive jet machining.

Nozzles made of tungsten carbide have an average life of 10 to 20 hours. While nozzles of sapphire last for about 300 hours of operation when used with 27 NS abrasive powder. The gases used are nitrogen carbon dioxide or clear air.

The metal removal rate depends upon the diameter of nozzle, composition of abrasive gas mixture, set pressure, hardness of abrasive particles and that of work material, particle size, velocity jet and distance of work piece from the jet. A typical material removal rate for abrasive jet machining is 16 mm^3/min in cutting glass.

Advantage:

- AJM process is a highly flexible process wherein the abrasive media is carried by a flexible hose, which can reach out to some difficult areas and internal regions.

- AJM process creates localized forces and generates lesser heat than the conventional machining processes.

- There is no damage to the work piece surface and also the process does not have tool-work piece contact, hence lesser amount of heat is generated.

- The power consumption in AJM process is low.

Disadvantages:

- The material removal rate is low.

- The process is limited to brittle and hard materials.

- The wear rate of nozzle is very high.

- The process results in poor machining accuracy.

- The process can cause environmental pollution.

Applications

Metal working:

- De-burring of some critical zones in the machined parts.

- Drilling and cutting of the thin and hardened metal sections.

- Removing the machining marks, flaws, chrome and anodizing marks.

Glass:

- Cutting of the optical fibers without altering its wavelength.

- Cutting, drilling and frosting precision optical lenses.

- Cutting extremely thin sections of glass and intricate curved patterns.

- Cutting and etching normally inaccessible areas and internal surfaces.

- Cleaning and dressing the grinding wheels used for glass.

Grinding:

- Cleaning the residues from diamond wheels, dressing wheels of any shape and size.

8.2.1 Water Jet Machining (WJM)

Water jet machine is a very clean job with no dust or noise, odor and very little, in fact, the health and safety and environmental impact is minimal, almost. Nozzles can be installed in multi-axis robotic arm cutting complex shapes in three-dimensional reality, this setting has been employed successfully in car dashboards, cut out foam laminate.

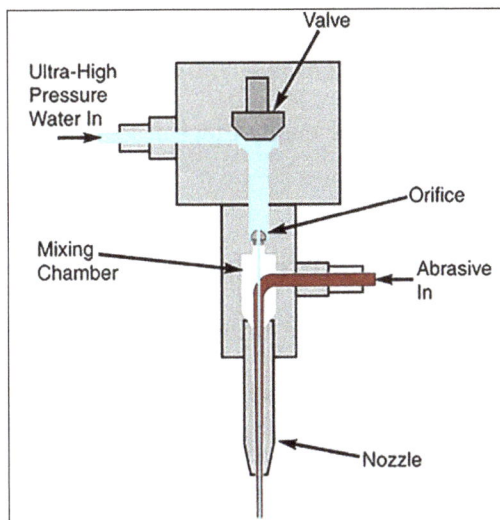

Water jet machining.

To cut it harder and more resistant materials such as metals, abrasive particles such as garnet or alumina is added to water prior to entering the cutting zone. This is called a water polishing Jet Machining (AWJ).

Pressure up to about 1400 MPa - 4000 MPa even more than usual for the effective functioning abrasive solution is forced through into the head between 0.05mm and 1mm diameter at a rate of 0.5 liter to 25 liters per minute, which resulted in jet speed in the region from 520 to 914 meters per second.

This serves as the blade, eroding many materials quickly In fact, due to the continuous change contrary to the cutting edge, fresh particles tend to be presented to the work piece, so they are not dull or stiff but flexible material being cut. The nozzle orifices of the materials made generally very difficult to man-made Ruby and Sapphire, as well as carbide composites are used .Similar to the laser to cut the maximum thickness of a majority interest in the area of 25 mm.

Advantages of Water Jet Machining (WJM):

- Water jet machining is a relatively fast process.
- It prevents the formation of heat affected zones on the work piece.
- It automatically cleans the surface of the work piece.
- WJM has excellent precision. Tolerances of the order of ±0.005″ can be obtained.
- It does not produce any hazardous gas.
- It is eco-friendly.

Disadvantages of Water Jet Machining:

- Only soft materials can be machined.
- Very thick materials cannot be easily machined.
- Initial investment is high.

Applications of Water Jet Machining

- Water jet machining is used to cut thin non-metallic sheets.
- It is used to cut rubber, wood, ceramics and many other soft materials.
- It is used for machining circuit boards.
- It is used in food industry.

8.3   Electron Beam Machining

Electron Beam Machining (EBM) is a thermal process. Here a steam of high speed electrons impinges on the work surface so that the kinetic energy of electrons is transferred to work producing intense heating.

Depending upon the intensity of heating the work piece can melt and vaporize. The process of heating by electron beam is used for annealing, welding or metal removal.

During EBM process very high velocities can be obtained by using enough voltage of 1,50,000 V which can produce velocity of 228,478 km/sec and it is focused on 10 – 200 μm diameter. Power density can go up to 6500 billion W/sq.mm. Such a power density can vaporize any substance immediately.

Complex contours can be easily machined by maneuvering the electron beam using magnetic deflection coils.

To avoid a collision of the accelerating electrons with the air molecules, the process has to be conducted in vacuum. So EBM is not suitable for large work pieces. Process is accomplished with vacuum so no possibility of contamination. No effects on work piece because about 25-50μm away from machining spot remains at room temperature and so no effects of high temperature on work.

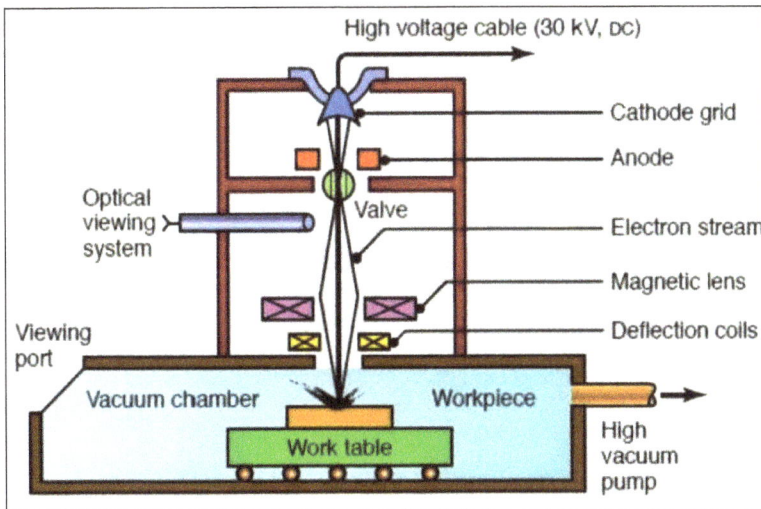

Electron Beam Machining.

Advantages:

- It can produced very small size holes.

- Surface finish produced is good.

- Highly reactive metals like Al and Mg can be machined very easily.

Limitations:

- Material removal rate is very low compared to other convectional machining processes.

- Maintaining perfect vacuum is very difficult.

- The machining process can't be seen by operator.

- Work piece material should be electrically conducting.

Applications

- Used for producing very small size holes like holes in diesel injection nozzles, Air brakes etc.

- Used only for circular holes.

8.3.1 Electron Discharge Machining

Electrical Discharge Machining (EDM), also known as spark erosion, employs electrical energy to remove metal from the work piece without touching it.

A pulsating high- frequency electric current is applied between the tool point and the work piece, causing sparks to jump the gap and vaporize small areas of the work piece. Because no cutting forces are involved, light, delicate operations can be performed on thin work pieces.

EDM can produce shapes unobtainable by any conventional machining process.

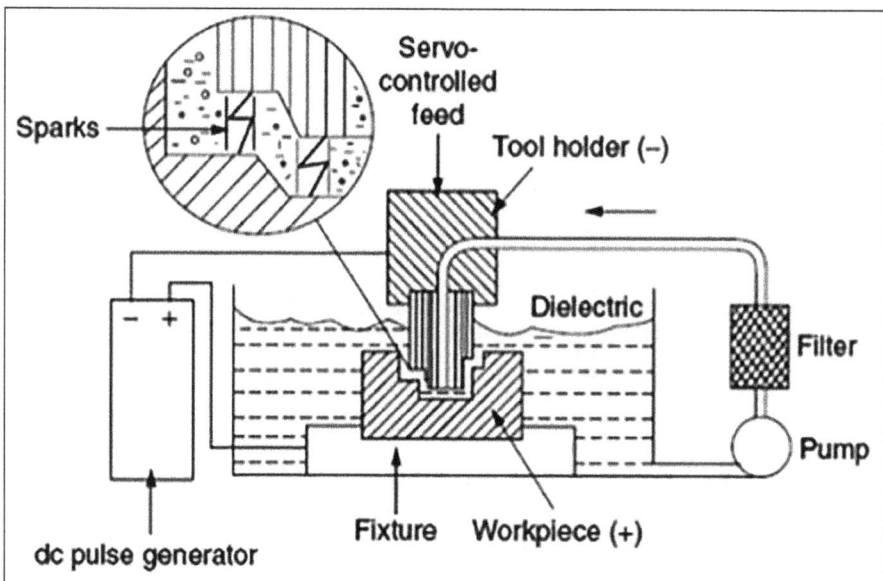

Electron discharging machining.

Ram EDM

A process using a shaped electrode made from graphite or copper. The electrode is separated by a nonconductive liquid and maintained at a close distance (about 0.001").

A high DC voltage is pulsed to the electrode and jumps to the conductive work piece. The resulting sparks erode the work piece and generate a cavity in the reverse shape of the electrode or a through hole in the case of a plain electrode.

Permits machining shapes to tight accuracies without the internal stresses conventional machining often generates. Also known as "die-sinker" or "sinker" electrical-discharge machining.

Wire EDMA process similar to sinker electrical-discharge machining except a small-diameter copper or brass wire is used as a traveling electrode.

The process is usually used in conjunction with a CNC and will only work when a part is to be cut completely through. A common analogy is to describe wire electrical-discharge machining as an ultraprecise, electrical, contour-sawing operation.

Advantages of EDM:

- Complex shapes that would otherwise be difficult to produce with conventional cutting tools.
- Extremely hard material to very close tolerances.
- Very small work pieces where conventional machining tools may damage the part from excess cutting tool pressure.
- There is no direct contact between tool and work piece. Therefore delicate sections and weak materials can be machined without any distortion.

Disadvantages of EDM:

- Relatively low rate of material removal.
- Additional lead time and cost used for creating electrodes for ram/sinker EDM.
- Reproducing sharp corners on the work piece is difficult due to electrode wear.
- Electrical power consumption is high.
- Material must be electrically conductive.

Applications

EDM permits machining shapes to tight accuracies without the internal stresses conventional machining often generates. Useful in die-making.

Permissions

All chapters in this book are published with permission under the Creative Commons Attribution Share Alike License or equivalent. Every chapter published in this book has been scrutinized by our experts. Their significance has been extensively debated. The topics covered herein carry significant information for a comprehensive understanding. They may even be implemented as practical applications or may be referred to as a beginning point for further studies.

We would like to thank the editorial team for lending their expertise to make the book truly unique. They have played a crucial role in the development of this book. Without their invaluable contributions this book wouldn't have been possible. They have made vital efforts to compile up to date information on the varied aspects of this subject to make this book a valuable addition to the collection of many professionals and students.

This book was conceptualized with the vision of imparting up-to-date and integrated information in this field. To ensure the same, a matchless editorial board was set up. Every individual on the board went through rigorous rounds of assessment to prove their worth. After which they invested a large part of their time researching and compiling the most relevant data for our readers.

The editorial board has been involved in producing this book since its inception. They have spent rigorous hours researching and exploring the diverse topics which have resulted in the successful publishing of this book. They have passed on their knowledge of decades through this book. To expedite this challenging task, the publisher supported the team at every step. A small team of assistant editors was also appointed to further simplify the editing procedure and attain best results for the readers.

Apart from the editorial board, the designing team has also invested a significant amount of their time in understanding the subject and creating the most relevant covers. They scrutinized every image to scout for the most suitable representation of the subject and create an appropriate cover for the book.

The publishing team has been an ardent support to the editorial, designing and production team. Their endless efforts to recruit the best for this project, has resulted in the accomplishment of this book. They are a veteran in the field of academics and their pool of knowledge is as vast as their experience in printing. Their expertise and guidance has proved useful at every step. Their uncompromising quality standards have made this book an exceptional effort. Their encouragement from time to time has been an inspiration for everyone.

The publisher and the editorial board hope that this book will prove to be a valuable piece of knowledge for students, practitioners and scholars across the globe.

Index